Student Study Guide

Electric Circuits

Ninth Edition

Student Study Guide

Electric Circuits

Ninth Edition

James W. Nilsson
Professor Emeritus
Iowa State University

Susan A. Riedel
Marquette University

Prentice Hall

Boston · Columbus · Indianapolis · New York · San Francisco · Upper Saddle River
Amsterdam · Cape Town · Dubai · London · Madrid · Milan · Munich · Paris · Montreal · Toronto
Delhi · Mexico City · Sao Paulo · Sydney · Hong Kong · Seoul · Singapore · Taipei · Tokyo

Editorial Director, Computer Science and Engineering: *Marcia J. Horton*
Senior Editor: *Andrew Gilfillan*
Associate Editor: *Alice Dworkin*
Editorial Assistant: *William Opaluch*
Director of Marketing: *Margaret Waples*
Marketing Manager: *Tim Galligan*
Vice President, Production: *Vince O'Brien*
Senior Managing Editor: *Scott Disanno*
Production Editor: *Irwin Zucker*
Art Director: *Jayne Conte*
Art Editor: *Gregory Dulles*
Senior Operations Specialist: *Alan Fischer*
Marketing Assistant: *Mack Patterson*

Prentice Hall
is an imprint of

PEARSON

www.pearsonhighered.com

10 9 8 7 6 5 4 3 2 1

ISBN-13: 978-0-13-213218-3
ISBN-10:   0-13-213218-4

Contents

Student Study Guide

Electric Circuits

Ninth Edition

Chapter 1

Balancing Power in DC Circuits

Balancing power in dc circuits is a good method for verifying that the voltages and currents calculated for the elements in the circuit are consistent. When the power balances, the sum of the powers for each circuit component will be zero. This means that all of the power generated in the circuit is absorbed by the circuit, so the net power is zero. We will use power balancing to confirm the voltages and currents calculated when using the node voltage and mesh current methods of circuit analysis illustrated in upcoming chapters. Performing a power balance depends on using the **passive sign convention** correctly. Remember that the passive sign convention tells us whether to use a positive or a negative sign in equations relating voltage and current for a single circuit component. The passive sign convention applies to the power equation as follows:

- When the arrow indicating current direction in a component points from the positive voltage polarity mark to the negative voltage polarity mark for that same component, $p = +vi$;

- When the arrow indicating current dirction in a component points from the negative voltage polarity mark to the positive voltage polarity mark for that same component, $p = -vi$.

In other words, when the current arrow points from + to −, use a + sign in the equation, and when the current arrow points from − to + use a − sign in the equation — the current arrow points to the sign to be used in the equation.

Performing a power balance consists of the following steps:

1. Create a table with five (5) columns. The first column will identify the component whose power is to be calculated. The second column is the value of the voltage drop across that component, from the + polarity mark to the − polarity mark. The third column is the value of the current flowing through that component, in the direction of the current arrow. The fourth column is the equation to be used to calculate the power for this component. In the examples, this equation will be $+vi$ or $-vi$, where the sign in the equation is determined by the passive sign convention. The fifth column is the numerical value of the power for that component, calculated by substituting the current and voltage values into the equation. Remember that when the power is greater than zero, the component is absorbing power from the circuit, and when the power is less than zero the component is delivering power to the circuit.

2. Identify each component in Column 1, and fill in that component's voltage and current in Columns 2 and 3, respectively.

3. Using the passive sign convention, determine whether the power equation for each component is $+vi$ or $-vi$ and fill in the equation in Column 4.

4. Substitute the value for voltage and current into the power equation and compute the power for each component. Put this value in Column 5.

5. When the table is completed, sum all of the power values in Column 5. If the sum is zero, the power balances for this circuit.

We begin with circuits whose power balances. Once you have balanced the power for these circuits, we move on to circuits whose power does not balance due to a sign error, and suggest a method for finding the error.

Example 1.1

Create a table to determine whether or not the power balances for the circuit in Fig. 1.1.

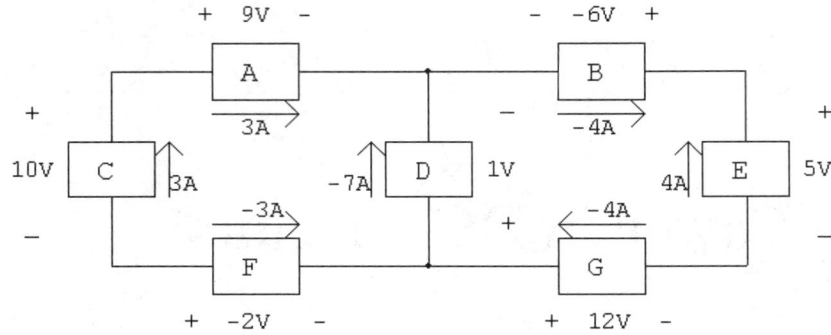

Figure 1.1: The circuit for Example 1.1

Solution

1. The five column table is shown below, with the columns labeled. We have seven rows in the table to accommodate the seven circuit components in Fig. 1.1.

Component	v (V)	i (A)	Equation	p (W)

2. We fill in Columns 1, 2, and 3 by identifying the component in Column 1, copying its voltage from Fig. 1.1 into Column 2, and copying its current from Fig. 1.1 into Column 3. The result is the partially completed table shown below:

Component	v (V)	i (A)	Equation	p (W)
A	9	3		
B	−6	−4		
C	10	3		
D	1	−7		
E	5	4		
F	−2	−3		
G	12	−4		

3. Now, pay careful attention to the voltage polarity and current direction for each component, and use this information together with the passive sign convention to determine whether the power equation for each component is $+vi$ or $-vi$. Visualize the current arrow aligned with the voltage polarity markings; then, the current arrow points to the correct sign. The result is the partially completed table shown below:

Component	v (V)	i (A)	Equation	p (W)
A	9	3	$+vi$	
B	−6	−4	$-vi$	
C	10	3	$-vi$	
D	1	−7	$+vi$	
E	5	4	$-vi$	
F	−2	−3	$+vi$	
G	12	−4	$-vi$	

4. Now substitute the values for voltage and current from Columns 2 and 3 into the equation in Column 4, paying close attention to all signs. The resulting value for the power should be placed in Column 5. The completed table is shown below:

Component	v (V)	i (A)	Equation	p (W)
A	9	3	$+vi$	27
B	−6	−4	$-vi$	−24
C	10	3	$-vi$	−30
D	1	−7	$+vi$	−7
E	5	4	$-vi$	−20
F	−2	−3	$+vi$	6
G	12	−4	$-vi$	48

5. Finally, we use the completed table to determine whether the power is balanced by summing the power values in Column 5:

$$27 + (-24) + (-30) + (-7) + (-20) + 6 + 48 = 0 \text{ W}$$

As we expected, the power balances.

Now try using the power balance method for the practice problems below.

Practice Problem 1.1

Determine whether or not the power balances for the circuit in Fig. 1.2.

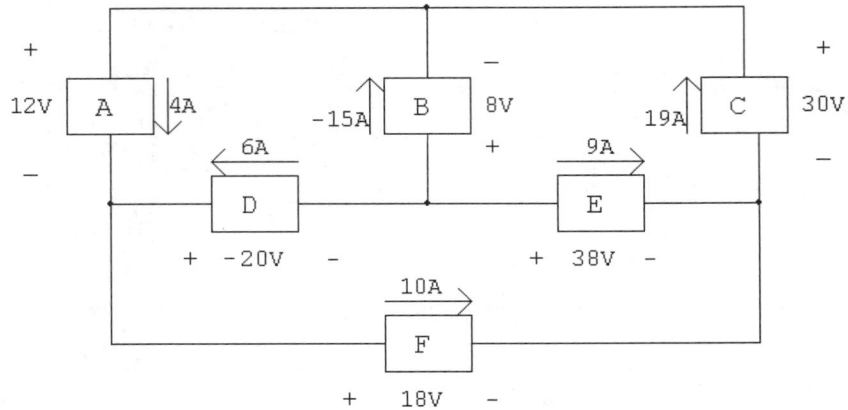

Figure 1.2: The circuit for Practice Problem 1.1

1. The table is shown below, with columns labeled and six rows to accommodate the six components in Fig.1.2.

2. Fill in Columns 1, 2, and 3 by identifying the component in Column 1, copying its voltage from Fig. 1.2 into Column 2, and copying its current from Fig. 1.2 into Column 3.

3. Determine whether the power equation for each component is $+vi$ or $-vi$. Place the appropriate equation in Column 4.

4. Substitute the values for voltage and current from Columns 2 and 3 into the equation in Column 4 and place the resulting value for the power in Column 5 to complete the table.

5. Use the completed table to determine whether the power is balanced by summing the power values in Column 5.

Component	v (V)	i (A)	Equation	p (W)

Practice Problem 1.2

Determine whether or not the power balances for the circuit in Fig. 1.3.

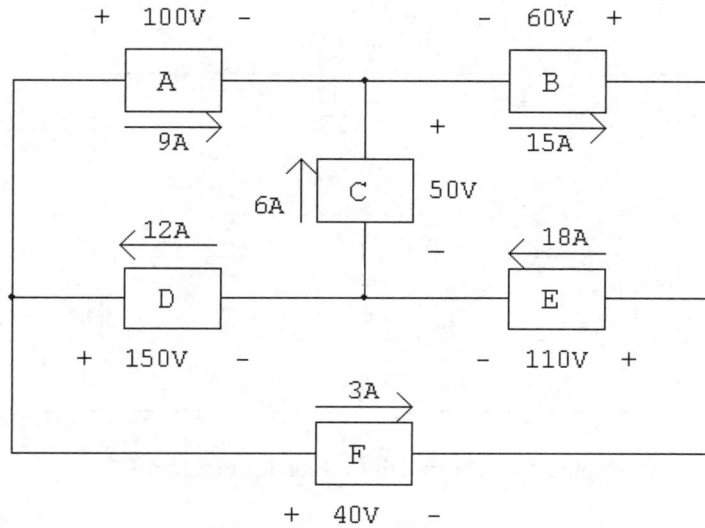

Figure 1.3: The circuit for Practice Problem 1.2

1. The table is shown below, with columns labeled and six rows to accommodate the six components in Fig.1.3.

2. Fill in Columns 1, 2, and 3 by identifying the component in Column 1, copying its voltage from Fig. 1.3 into Column 2, and copying its current from Fig. 1.3 into Column 3.

3. Determine whether the power equation for each component is $+vi$ or $-vi$. Place the appropriate equation in Column 4.

4. Substitute the values for voltage and current from Columns 2 and 3 into the equation in Column 4 and place the resulting value for the power in Column 5 to complete the table.

5. Use the completed table to determine whether the power is balanced by summing the power values in Column 5.

Component	v (V)	i (A)	Equation	p (W)

Practice Problem 1.3

Determine whether or not the power balances for the circuit in Fig. 1.4.

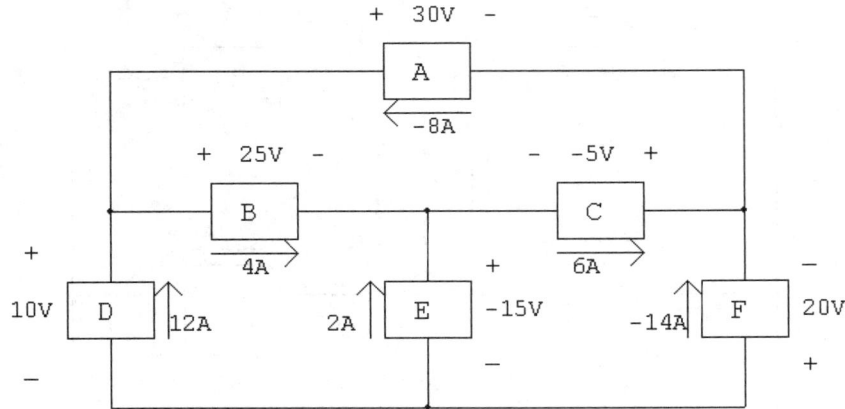

Figure 1.4: The circuit for Practice Problem 1.3

1. The table is shown below, with columns labeled and six rows to accommodate the six components in Fig.1.4.

2. Fill in Columns 1, 2, and 3 by identifying the component in Column 1, copying its voltage from Fig. 1.4 into Column 2, and copying its current from Fig. 1.4 into Column 3.

3. Determine whether the power equation for each component is $+vi$ or $-vi$. Place the appropriate equation in Column 4.

4. Substitute the values for voltage and current from Columns 2 and 3 into the equation in Column 4 and place the resulting value for the power in Column 5 to complete the table.

5. Use the completed table to determine whether the power is balanced by summing the power values in Column 5.

Component	v (V)	i (A)	Equation	p (W)

Example 1.2

We know that one sign error exists in the circuit in Fig. 1.5. Create a table to show that the power does not balance. Use the table to determine where the error exists, and correct the error.

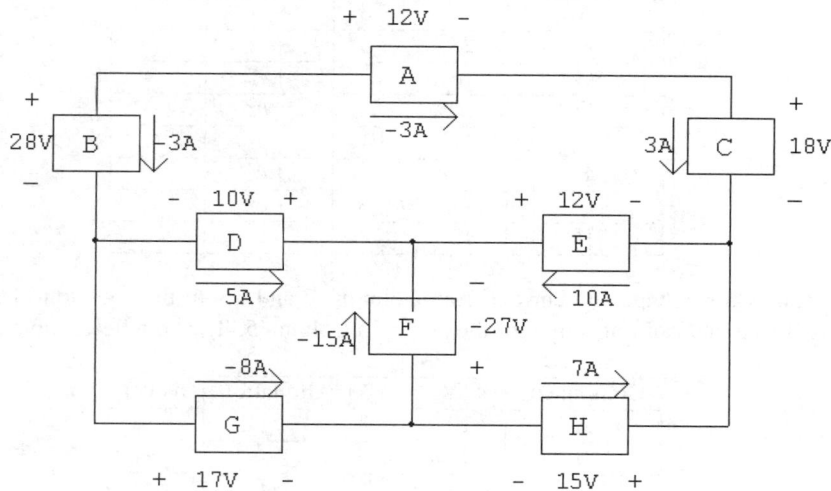

Figure 1.5: The circuit for Example 1.2

Solution

1. The five column table is shown below, with the columns labeled. We have eight rows in the table to accommodate the eight circuit components in Fig. 1.5.

Component	v (V)	i (A)	Equation	p (W)

2. We fill in Columns 1, 2, and 3 by identifying the component in Column 1, copying its voltage from Fig. 1.5 into Column 2, and copying its current from Fig. 1.5 into Column 3. The result is the partially completed table shown below:

Component	v (V)	i (A)	Equation	p (W)
A	12	−3		
B	28	−3		
C	18	3		
D	10	5		
E	12	10		
F	−27	−15		
G	17	−8		
H	15	7		

3. Now, pay careful attention to the voltage polarity and current direction for each component, and use this information together with the passive sign convention to determine whether the power equation for each component is $+vi$ or $-vi$. The result is the partially completed table shown below:

Component	v (V)	i (A)	Equation	p (W)
A	12	−3	$+vi$	
B	28	−3	$+vi$	
C	18	3	$+vi$	
D	10	5	$-vi$	
E	12	10	$-vi$	
F	−27	−15	$+vi$	
G	17	−8	$+vi$	
H	15	7	$-vi$	

4. Now substitute the values for voltage and current from Columns 2 and 3 into the equation in Column 4, paying close attention to all signs. Place the resulting value for the power in Column 5. The completed table is shown below:

Component	v (V)	i (A)	Equation	p (W)
A	12	−3	$+vi$	−36
B	28	−3	$+vi$	−84
C	18	3	$+vi$	54
D	10	5	$-vi$	−50
E	12	10	$-vi$	−120
F	−27	−15	$+vi$	405
G	17	−8	$+vi$	−136
H	15	7	$-vi$	−105

5. Use the completed table to determine whether the power is balanced by summing the power values in Column 5:

$$(-36) + (-84) + 54 + (-50) + (-120) + 405 + (-136) + (-105) = -72 \text{ W}$$

The power does not balance, because there is a sign error in Fig. 1.5. One way to find the sign error is to divide the sum of the power values by 2. This works because a single sign error causes a power value to have a positive sign instead of a negative sign (or vice versa), so the power is added instead of subtracted (or vice versa) and the sum of the power values is then twice the power whose sign is in error. In this example, half the sum of the power values is $-72/2 = -36$ W, which is the power associated with component A. Looking at Fig. 1.5 it is easy to see that the current value assigned to component A should be $+3A$, to $-3A$. This makes the current in component A consistent with the currents assigned to components B and C which are in series with component A and must have the same current. Thus, the power for component A is $+36$W; when we sum the power values we get

$$36 + (-84) + 54 + (-50) + (-120) + 405 + (-136) + (-105) = 0 \text{ W}$$

so now the power balances.

Now try using the power balance method for the practice problems below, each of which contains a single sign error. When the power does not balance, identify the component containing the sign error and correct the error.

Practice Problem 1.4

Determine whether or not the power balances for the circuit in Fig. 1.6.

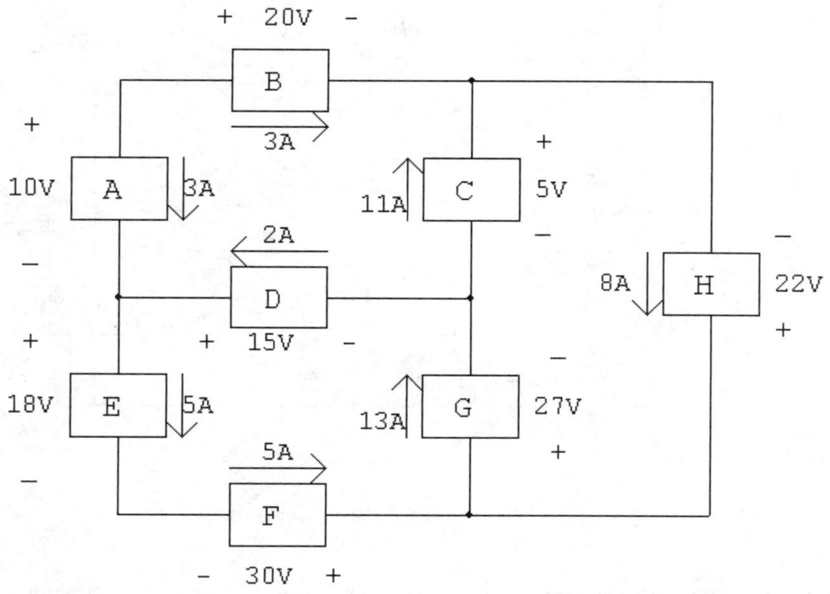

Figure 1.6: The circuit for Practice Problem 1.4

1. The table is shown below, with columns labeled and eight rows to accommodate the eight components in Fig.1.6.

2. Fill in Columns 1, 2, and 3 by identifying the component in Column 1, copying its voltage from Fig. 1.6 into Column 2, and copying its current from Fig. 1.6 into Column 3.

3. Determine whether the power equation for each component is $+vi$ or $-vi$. Place the appropriate equation in Column 4.

4. Substitute the values for voltage and current from Columns 2 and 3 into the equation in Column 4 and place the resulting value for the power in Column 5 to complete the table.

5. Use the completed table to determine whether the power is balanced by summing the power values in Column 5. If the power is not balanced, determine which component has a sign error and correct the error.

Component	v (V)	i (A)	Equation	p (W)

Practice Problem 1.5

Determine whether or not the power balances for the circuit in Fig. 1.7.

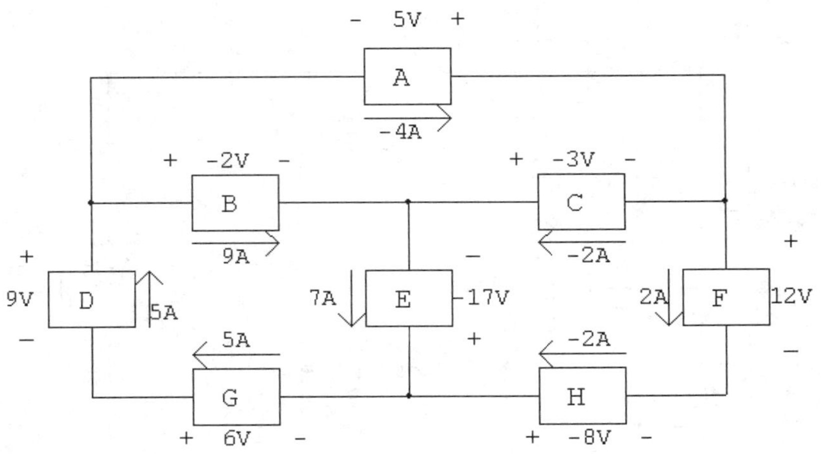

Figure 1.7: The circuit for Practice Problem 1.5

1. The table is shown below, with columns labeled and eight rows to accommodate the eight components in Fig.1.7.

2. Fill in Columns 1, 2, and 3 by identifying the component in Column 1, copying its voltage from Fig. 1.7 into Column 2, and copying its current from Fig. 1.7 into Column 3.

3. Determine whether the power equation for each component is $+vi$ or $-vi$. Place the appropriate equation in Column 4.

4. Substitute the values for voltage and current from Columns 2 and 3 into the equation in Column 4 and place the resulting value for the power in Column 5 to complete the table.

5. Use the completed table to determine whether the power is balanced by summing the power values in Column 5. If the power is not balanced, determine which component has a sign error and correct the error.

Component	v (V)	i (A)	Equation	p (W)

Practice Problem 1.6

Determine whether or not the power balances for the circuit in Fig. 1.8.

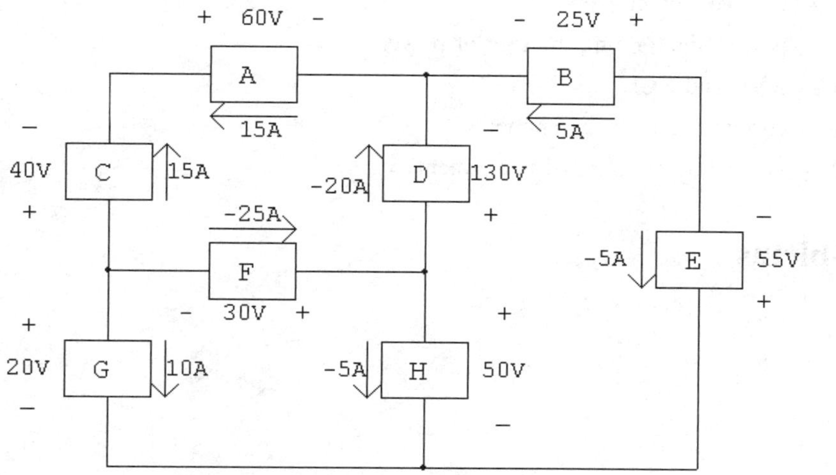

Figure 1.8: The circuit for Practice Problem 1.6

1. The table is shown below, with columns labeled and eight rows to accommodate the eight components in Fig.1.8.

2. Fill in Columns 1, 2, and 3 by identifying the component in Column 1, copying its voltage from Fig. 1.8 into Column 2, and copying its current from Fig. 1.8 into Column 3.

3. Determine whether the power equation for each component is $+vi$ or $-vi$. Place the appropriate equation in Column 4.

4. Substitute the values for voltage and current from Columns 2 and 3 into the equation in Column 4 and place the resulting value for the power in Column 5 to complete the table.

5. Use the completed table to determine whether the power is balanced by summing the power values in Column 5. If the power is not balanced, determine which component has a sign error and correct the error.

Component	v (V)	i (A)	Equation	p (W)

Reading

- *Electrical Circuits*, ninth edition

    ◆ Section 1.5 — passive sign convention
    ◆ Section 1.6 — power; interpreting the sign of power
    ◆ Section 2.4 — KVL and KCL
    ◆ Section 3.1 — currents in series components
    ◆ Section 3.2 — voltage across parallel components

Additional Problems

- 1.26 — 1.30

Solutions

- Practice Problem 1.1:

$$48 - 120 - 570 + 120 + 342 + 180 = 0\text{W}$$

- Practice Problem 1.2:

$$900 - 900 - 300 - 1800 + 1980 + 120 = 0\text{W}$$

- Practice Problem 1.3:

$$240 + 100 + 30 - 120 + 30 - 280 = 0\text{W}$$

- Practice Problem 1.4:

$$30 + 60 - 55 - 30 + 90 - 150 + 351 - 176 = 120\text{W}$$

The current in component B should be -3A.

- Practice Problem 1.5:

$$20 - 18 - 6 - 45 + 119 + 24 - 30 - 16 = 48\text{W}$$

The current in component F should be -2A.

- Practice Problem 1.6:

$$-900 + 125 + 600 - 2600 + 275 + 750 + 200 - 250 = -1800\text{W}$$

The current in component A should be -15A.

Chapter 2

Combining Resistors in Series and in Parallel

We can combine resistors in series and in parallel to reduce the number of resistors in a circuit, which often simplifies any analysis we wish to perform on the circuit. Resistors are **in series** when they are connected end-to-end and have exactly the same current. Resistors are **in parallel** when they are connected at two points and have exactly the same voltage drop. Be sure to check your circuit carefully to be sure that the conditions are satisfied for series or parallel connections before combining resistors.

The best way to combine resistors in series and in parallel is to make the combinations one step at a time, redrawing the circuit after each step. You are less likely to make a mistake if you take a methodical approach and take the time to redraw the circuit. We propose a four step method, as follows:

1. Draw the simplified circuit you expect to construct once all appropriate resistors are combined in series and in parallel. This simplified circuit should contain any components whose voltages or currents were sought in the original circuit.

2. Starting from the original circuit, make a combinations of resistors in series or in parallel, one step at a time. Redraw the circuit after each step. Make sure that any components whose voltages or currents were sought in the original circuit remain in each of your drawings. The last combination you make should result in a circuit that looks like the simplified circuit from Step 1.

3. Use the simplified circuit to calculate any voltages or currents specified in the original circuit. Usually these calculations involve a simple circuit analysis technique, such as Ohm's law, current division, or voltage division.

4. Check your solution. To do this, redraw the original circuit and label it with the voltages or currents you calculated in Step 3. Use these values to calculate the voltages or currents for the remaining components in the circuit. Finally, calculate the power for each component and confirm that the power balances.

We illustrate this method with the two examples that follow.

Example 2.1

Combine resistors in series and in parallel to simplify the circuit in Fig. 2.1 so it is easy to calculate the current i.

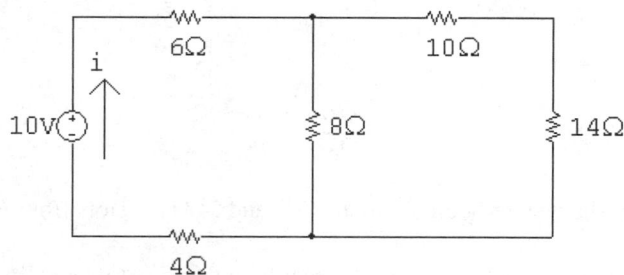

Figure 2.1: The circuit for Example 2.1

Solution

1. We want to calculate the current i flowing in the voltage source in Fig. 2.1. Therefore we can combine all of the resistors in this figure into one equivalent resistor, as seen in Fig. 2.2.

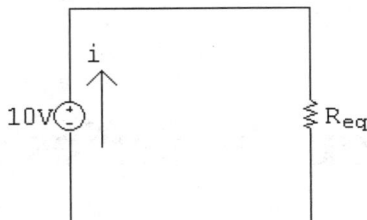

Figure 2.2: The circuit for Eaxmple 2.1, with all resistors combined into one equivalent.

2. It is usually best to combine the resistors working from one side of the circuit to the other. We will start on the right side of this circuit, as that is the furthest from the location of the current we wish to calculate. We see that the $10\,\Omega$ resistor and the $14\,\Omega$ resistor are connected end-to-end, so have the same current flowing through them. Thus they are in series, so they can be combined into a single resistor whose value is $10 + 14 = 24\,\Omega$. The resulting simplified circuit is shown in Fig. 2.3.

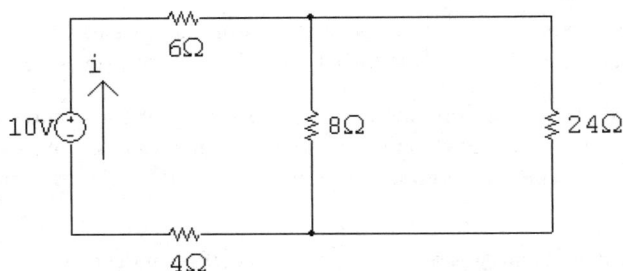

Figure 2.3: The result of combining the $10\,\Omega$ and $14\,\Omega$ resistors from Fig. 2.1.

Contining from the right side of Fig. 2.3 we see that the $8\,\Omega$ resistor and the $24\,\Omega$ resistor are connected at both ends, so have the same voltage drop across them. Thus they are in parallel, so can be combined into a single resistor whose value is $(8)(24)/(8+24) = 6\,\Omega$. The resulting simplified circuit is shown in Fig. 2.4.

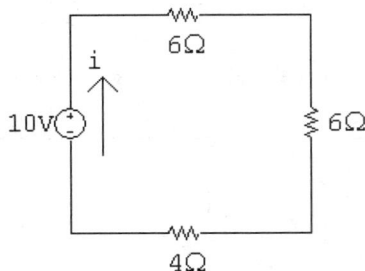

Figure 2.4: The result of combining the $8\,\Omega$ and $24\,\Omega$ resistors from Fig. 2.3.

Continuing with the simplified circuit in Fig. 2.4 we see that both $6\,\Omega$ resistors and the $4\,\Omega$ resistor are connected end-to-end, so have the same current flowing through them. This means that these three resistors are in series and can be combined into one equivalent resistor whose value is $6 + 6 + 4 = 16\,\Omega$. The resulting simplified circuit is shown in Fig. 2.5.

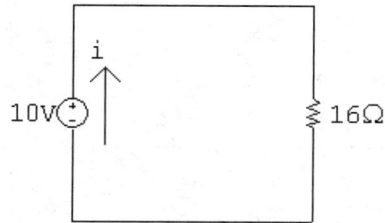

Figure 2.5: The result of combining all of the resistors from Fig. 2.4.

Note that Fig. 2.5 is exactly the same as Fig. 2.2 and that we now have calculated the equivalent resistance seen by the voltage source.

3. We can easily calculate the current in the voltage source in Fig. 2.4 using Ohm's law:

$$i = 10V/16\,\Omega = 625 \text{ mA}$$

4. We check our result by redrawing the original circuit with the current drawn from the voltage source indicated, as seen in Fig. 2.6.

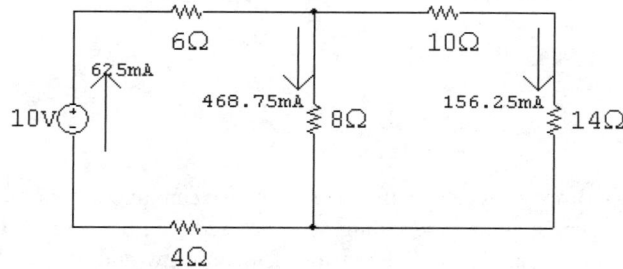

Figure 2.6: The original circuit with all currents calculated, ready to calculate power.

Using this current and current division, we can calculate the currents in all remaining components, also shown in Fig. 2.6:

$$
\begin{aligned}
i_{6\,\Omega} &= & i & = & 625 \text{ mA (in series with the voltage source)}; \\
i_{4\,\Omega} &= & i & = & 625 \text{ mA (in series with the voltage source)}; \\
i_{8\,\Omega} &= & (24)(0.625)/(32) & = & 468.75 \text{ mA (current division)}; \\
i_{10\,\Omega} &= & 625 - 468.75 & = & 156.25 \text{ mA (KCL)}; \\
i_{14\,\Omega} &= & i_{10\,\Omega} & = & 156.25 \text{ mA (in series with the 10\,\Omega resistor)};
\end{aligned}
$$

Now we can use these currents to calculate the power for each element:

$$
\begin{aligned}
p_{10V} &= & -vi &= & (10)(0.625)) &= & -6250 \text{ mW}; \\
p_{6\,\Omega} &= & i^2R &= & 0.625^2(6) &= & 2343.75 \text{ mW}; \\
p_{4\,\Omega} &= & i^2R &= & 0.625^2(4) &= & 1562.5 \text{ mW}; \\
p_{8\,\Omega} &= & i^2R &= & 0.46875^2(8) &= & 1757.813 \text{ mW}; \\
p_{10\,\Omega} &= & i^2R &= & 0.15625^2(10) &= & 244.141 \text{ mW}; \\
p_{14\,\Omega} &= & i^2R &= & 0.15625^2(14) &= & 341.797 \text{ mW};
\end{aligned}
$$

The sum of the powers is zero, confirming that our original calculation for the current i was correct.

Let's consider one more example before practicing these resistor combination techniques.

Example 2.2

Combine resistors in series and in parallel to simplify the circuit in Fig. 2.7 so it is easy to calculate the voltage v.

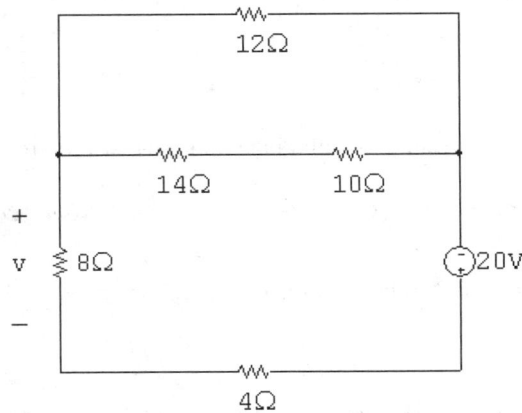

Figure 2.7: The circuit for Eaxmple 2.2

Solution

1. We want to calculate the voltage drop across a specific resistor, so we cannot combine this resistor with any others. If we do, we will lose the component across which the voltage v is defined. If possible, we would like to combine all of the remaining resistors in the circuit into one equivalent resistor. The circuit would then look like the one shown in Fig. 2.8. This circuit is easily analyzed using voltage division to find the requested voltage v.

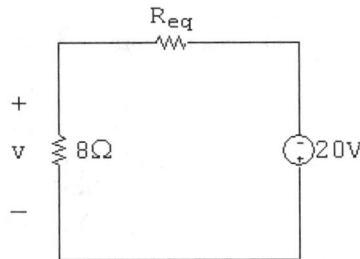

Figure 2.8: The circuit for Eaxmple 2.2, with all but one of the resistors combined into one equivalent, maintaining the resistor with the unknown voltage v.

2. Now we try to find resistors in series or in parallel. We see that the $10\,\Omega$ resistor and the $14\,\Omega$ resistor are connected end-to-end, so have the same current flowing through them. Thus they are in series, so they can be combined into a single resistor whose value is $10 + 14 = 24\,\Omega$. The resulting simplified circuit is shown in Fig. 2.9.

In Fig. 2.9 we see that the $12\,\Omega$ resistor and the $24\,\Omega$ resistor are connected at both ends, so have the same voltage drop across them. Thus they are in parallel, so can be combined into a single resistor whose value is $(12)(24)/(12+24) = 8\,\Omega$. The resulting simplified circuit is shown in Fig. 2.10.

Continuing with the simplified circuit in Fig. 2.10 we see that all of the components are connected end-to-end, and thus are in series. Any time components are in series we can rearrange their order without having any effect on the circuit. We make a rearrangement to place the $8\,\Omega$ resistor next to the $4\,\Omega$ resistor, as shown in Fig. 2.11.

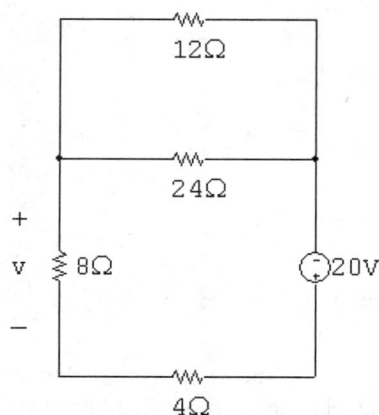

Figure 2.9: The result of combining the 10 Ω and 14 Ω resistors from Fig. 2.7.

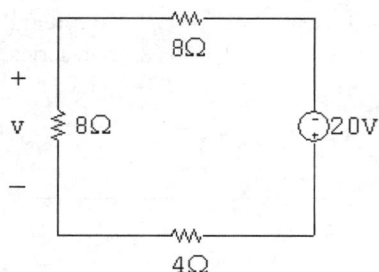

Figure 2.10: The result of combining the 12 Ω and 24 Ω resistors from Fig. 2.9.

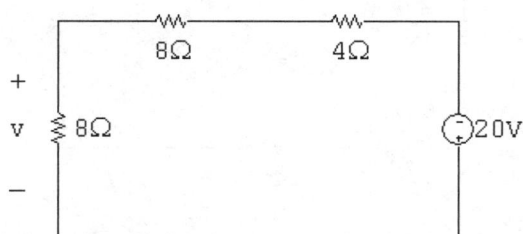

Figure 2.11: The result of rearranging the series-connected components in Fig. 2.10.

Remember that rearranging can also be performed with parallel connected components, as long as the parallel structure is maintained.

The final step is to combine the 8 Ω resistor and the 4 Ω resistor. The result is a $8 + 4 = 12\,\Omega$ resistor, shown in Fig. 2.12.

3. Now we can use voltage division to calculate the voltage v for the circuit in Fig. 2.12:

$$v = \frac{8\,\Omega}{8\,\Omega + 12\,\Omega}(-20V) = -8 \text{ V}$$

4. We check our result by redrawing the original circuit with the current drawn from the voltage source indicated, as seen in Fig. 2.13.

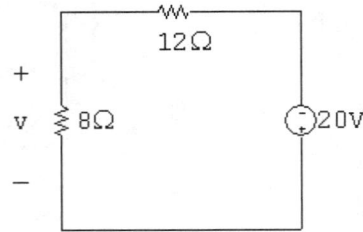

Figure 2.12: The result of combining the $8\,\Omega$ and $4\,\Omega$ resistors in Fig. 2.11.

Using this voltage we can calculate the current in the $8\,\Omega$ resistor and in the voltage source and $4\,\Omega$ resistor in series with it. We can then use current division to calculate the currents in all remaining components, also shown in Fig. 2.13:

$$
\begin{aligned}
i_{8\,\Omega} &= & -8/8 & = & -1 \text{ A (Ohm's law)};\\
i_{20\text{V}} &= & i_{8\,\Omega} & = & -1 \text{ A (in series with the } 8\,\Omega \text{ resistor)};\\
i_{4\,\Omega} &= & i_{8\,\Omega} & = & -1 \text{ A (in series with the } 8\,\Omega \text{ resistor)};\\
i_{14\,\Omega} &= & (12)(-1)/(36) & = & -1/3 \text{ A (current division)};\\
i_{10\,\Omega} &= & i_{14\,\Omega} & = & -1/3 \text{ A (in series with the } 14\,\Omega \text{ resistor)};\\
i_{12\,\Omega} &= & -1+1/3 & = & -2/3 \text{ A (KCL)};
\end{aligned}
$$

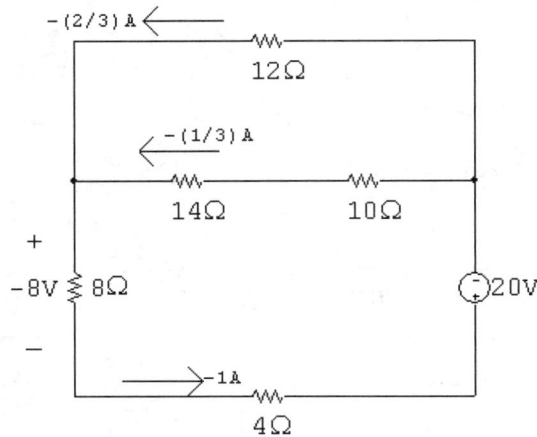

Figure 2.13: The original circuit with all currents calculated, ready to calculate power.

Now we can use these currents to calculate the power for each element:

$$
\begin{aligned}
p_{24\text{V}} &= & vi & = & (20)(-1)) & = & -20 \text{ W};\\
p_{8\,\Omega} &= & i^2 R & = & 1^2(8) & = & 8 \text{ W};\\
p_{4\,\Omega} &= & i^2 R & = & 1^2(4) & = & 4 \text{ W};\\
p_{10\,\Omega} &= & i^2 R & = & (1/3)^2(10) & = & 10/9 \text{ W};\\
p_{14\,\Omega} &= & i^2 R & = & (1/3)^2(14) & = & 14/9 \text{ W};\\
p_{12\,\Omega} &= & i^2 R & = & (2/3)^2(12) & = & 48/9 \text{ W};
\end{aligned}
$$

The sum of the powers is zero, confirming that our original calculation for the current v was correct.

Now you are ready to practice these circuit simplification techniques in the problems that follow.

Practice Problem 2.1

Find the voltage v for the circuit in Fig. 2.14.

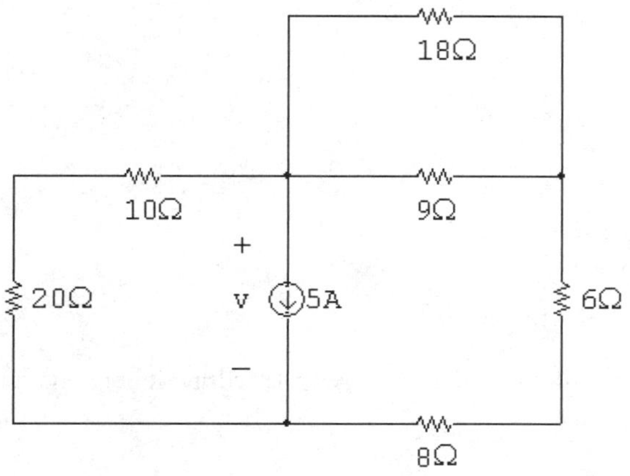

Figure 2.14: The circuit for Practice Problem 2.1.

1. In the space below, draw the simplified circuit that will result from series and parallel combinations of resistors. Be sure that the voltage v appears in this simplified circuit.

2. From the original circuit, combine resistors in series and in parallel, one step at a time, redrawing the circuit after each step. Remember that you can reorder components in series or components in parallel without affecting the circuit, if that helps you see further simplifications.

Continue the simplifications below.

3. Your final circuit should match the circuit you drew in Step 1; redraw it here and use it to calculate v.

4. Redraw the original circuit, placing the value of v in your drawing. Use that value of v to calculate currents for all other components. Then use these values to calculate power for each component, and confirm that the power balances.

Practice Problem 2.2

Find the voltage v for the circuit in Fig. 2.15.

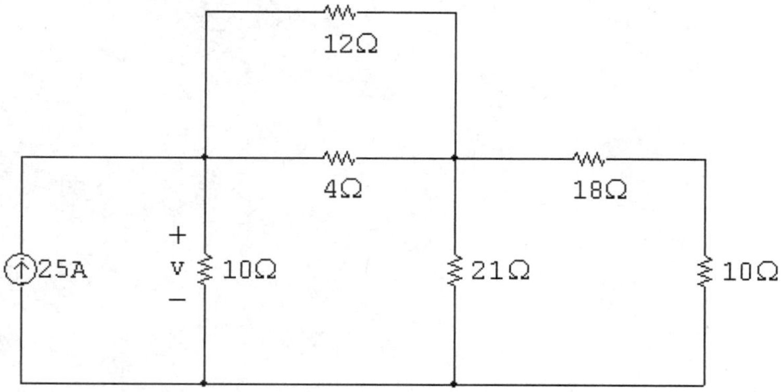

Figure 2.15: The circuit for Practice Problem 2.2.

1. In the space below, draw the simplified circuit that will result from series and parallel combinations of resistors. Be sure that the voltage v appears in this simplified circuit.

2. From the original circuit, combine resistors in series and in parallel, one step at a time, redrawing the circuit after each step. Remember that you can reorder components in series or components in parallel without affecting the circuit, if that helps you see further simplifications.

Continue the simplifications below.

3. Your final circuit should match the circuit you drew in Step 1; redraw it here and use it to calculate v.

4. Redraw the original circuit, placing the value of v in your drawing. Use that value of v to calculate currents for all other components. Then use these values to calculate power for each component, and confirm that the power balances.

Practice Problem 2.3

Find the current i for the circuit in Fig. 2.16.

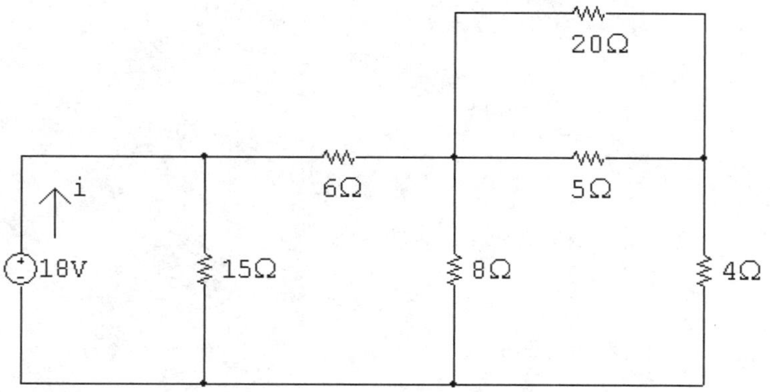

Figure 2.16: The circuit for Practice Problem 2.3.

1. In the space below, draw the simplified circuit that will result from series and parallel combinations of resistors. Be sure that the current i appears in this simplified circuit.

2. From the original circuit, combine resistors in series and in parallel, one step at a time, redrawing the circuit after each step. Remember that you can reorder components in series or components in parallel without affecting the circuit, if that helps you see further simplifications.

Continue the simplifications below.

3. Your final circuit should match the circuit you drew in Step 1; redraw it here and use it to calculate i.

4. Redraw the original circuit, placing the value of i in your drawing. Use that value of i to calculate currents for all other components. Then use these values to calculate power for each component, and confirm that the power balances.

Practice Problem 2.4

Find the current i for the circuit in Fig. 2.17.

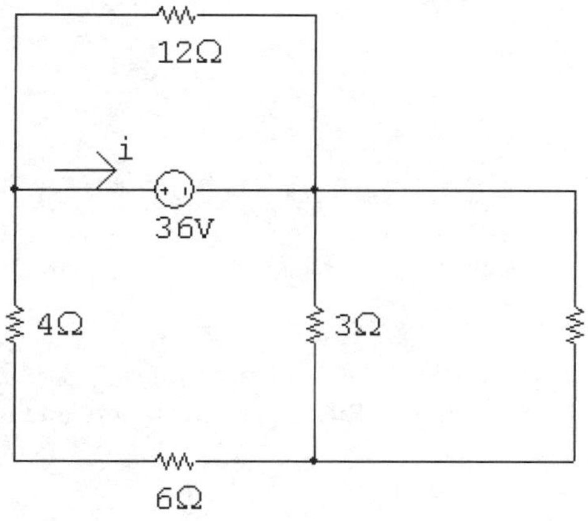

Figure 2.17: The circuit for Practice Problem 2.4.

1. In the space below, draw the simplified circuit that will result from series and parallel combinations of resistors. Be sure that the current i appears in this simplified circuit.

2. From the original circuit, combine resistors in series and in parallel, one step at a time, redrawing the circuit after each step. Remember that you can reorder components in series or components in parallel without affecting the circuit, if that helps you see further simplifications.

Continue the simplifications below.

3. Your final circuit should match the circuit you drew in Step 1; redraw it here and use it to calculate i.

4. Redraw the original circuit, placing the value of i in your drawing. Use that value of i to calculate currents for all other components. Then use these values to calculate power for each component, and confirm that the power balances.

Practice Problem 2.5

Find the voltage v for the circuit in Fig. 2.18.

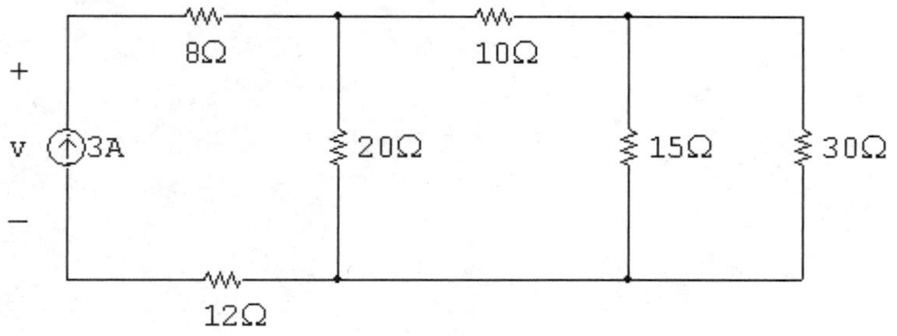

Figure 2.18: The circuit for Practice Problem 2.5.

1. In the space below, draw the simplified circuit that will result from series and parallel combinations of resistors. Be sure that the voltage v appears in this simplified circuit.

2. From the original circuit, combine resistors in series and in parallel, one step at a time, redrawing the circuit after each step. Remember that you can reorder components in series or components in parallel without affecting the circuit, if that helps you see further simplifications.

Continue the simplifications below.

3. Your final circuit should match the circuit you drew in Step 1; redraw it here and use it to calculate v.

4. Redraw the original circuit, placing the value of v in your drawing. Use that value of v to calculate currents for all other components. Then use these values to calculate power for each component, and confirm that the power balances.

Practice Problem 2.6

Find the current i for the circuit in Fig. 2.19.

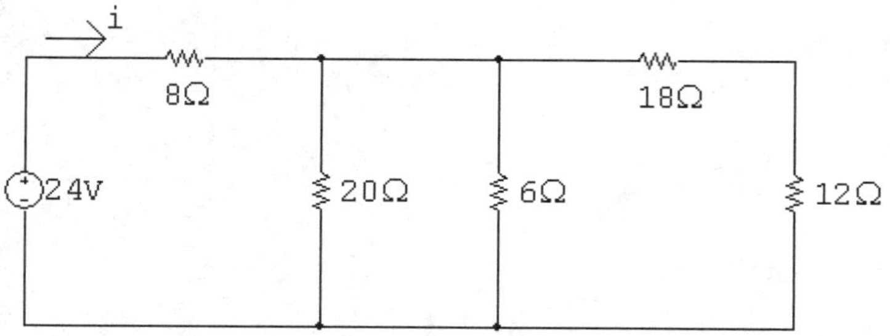

Figure 2.19: The circuit for Practice Problem 2.6.

1. In the space below, draw the simplified circuit that will result from series and parallel combinations of resistors. Be sure that the current i appears in this simplified circuit.

2. From the original circuit, combine resistors in series and in parallel, one step at a time, redrawing the circuit after each step. Remember that you can reorder components in series or components in parallel without affecting the circuit, if that helps you see further simplifications.

Continue the simplifications below.

3. Your final circuit should match the circuit you drew in Step 1; redraw it here and use it to calculate i.

4. Redraw the original circuit, placing the value of i in your drawing. Use that value of i to calculate currents for all other components. Then use these values to calculate power for each component, and confirm that the power balances.

Reading

- in *Electric Circuits*, ninth edition:

  ♦ Section 2.2 — Ohm's law

  ♦ Section 2.4 — KVL and KCL

  ♦ Section 3.1 — resistors in series

  ♦ Section 3.2 — resistors in parallel

  ♦ Section 3.3 — voltage and current dividers

  ♦ Section 3.4 — voltage and current division

- Workbook section — Balancing Power in DC Circuits

Additional Problems

- 3.1 – 3.7

- 3.13

Solutions

- Practice Problem 2.1:

$$R_{\text{eq}} = ((18\|9) + 6 + 8)\|(20 + 10) = 12\,\Omega \qquad v = 60 \text{ V}$$

- Practice Problem 2.2:

$$R_{\text{eq}} = \{[(18 + 10)\|21] + (12\|4)\}\|10 = 15\|10 = 6\,\Omega \qquad v = 150 \text{ V}$$

- Practice Problem 2.3:

$$R_{\text{eq}} = (((20\|5) + 4)\|8) + 6)\|15 = 6\,\Omega \qquad i = 3 \text{ A}$$

- Practice Problem 2.4:

$$R_{\text{eq}} = ((3\|6) + 4 + 6)\|12 = 6\,\Omega \qquad i = -6 \text{ A}$$

- Practice Problem 2.5:

$$R_{\text{eq}} = (((15\|30) + 10)\|20) + 12 + 8 = 30\,\Omega \qquad v = 90 \text{ V}$$

- Practice Problem 2.6:

$$R_{\text{eq}} = ((18 + 12)\|6\|20) + 8 = 12\,\Omega \qquad i = 2 \text{ A}$$

Chapter 3

Node Voltage Method

The node voltage method provides a systematic means to specify the equations needed to **solve a circuit**. The term "solve a circuit" means to find all of the voltages and all of the currents for all of the components in the circuit.

The node voltage method uses the **essential nodes** in a circuit. Remember that the essential nodes are the points in the circuit where three or more circuit elements are connected.

The node voltage method uses **KCL** equations that are written at certain essential nodes. Recall that KCL states that the sum of all of the currents at a node is zero.

The node voltage method is comprised of the following steps:

1. Identify all of the essential nodes in the circuit. To do this we will place a large black dot at each essential node.

2. Choose one of the essential nodes as the reference node. We will use a special symbol to label the reference node.

3. Assign variable names to each of the non-reference essential nodes. Each variable name represents the voltage drop between its node and the reference node. We will use variable names like v_1, v_a, v_Δ and so on.

4. Write a KCL equation at each of the non-reference essential nodes where the voltage with respect to the reference node is unknown. We will be methodical in writing these KCL equations, always summing the currents leaving the node.

5. Write any supplemental equations that are needed. These equations arise when there are dependent sources in the circuit, and when there are voltage sources in the circuit.

6. Express all of the equations in standard form. The standard form we use will allow our equations to be solved using a calculator, using a matrix method such as Cramer's rule, or using a computer tool such as MATLAB.

7. Solve the equations and check your solution using a power balance. If the power balances, use the solution to calculate the desired output value for the circuit.

We begin with an example that contains only resistors and independent current sources. Once you have mastered these types of circuits, we move on to example circuits containing dependent current sources, and then to circuits containing voltage sources.

Example 3.1

Using the node voltage method, find v_o for the circuit in Fig. 3.1.

Solution

1. Identify the essential nodes. There are four points at which three or more circuit elements connect, so there are four essential nodes. They have been labeled with large black dots in Fig. 3.2.

2. Chose a reference node. The choice of the reference node is entirely arbitrary; no matter which essential node is chosen, the voltages and currents that result from the analysis will have the same values. You should chose an essential node that makes the circuit analysis easier, if possible. In the circuit in Fig. 3.2 we have chosen the bottom node as the reference node, and labeled it with the symbol for circuit ground. We chose this node because it is one of the two nodes associated with v_o, the voltage of interest.

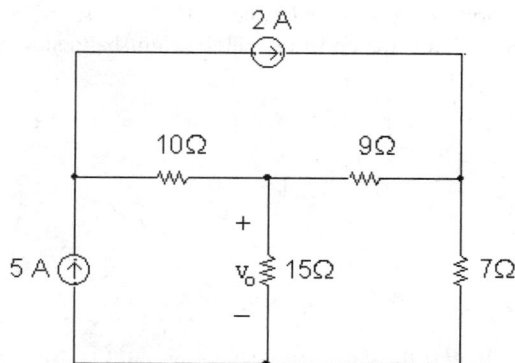

Figure 3.1: The circuit for Example 3.1

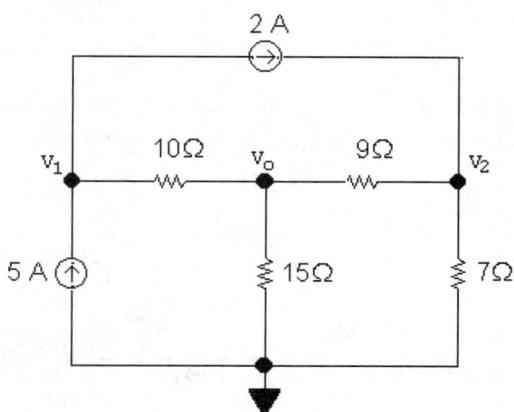

Figure 3.2: The circuit in Fig. 3.1 with the essential nodes marked, the reference node chosen, and the remaining essential nodes labeled.

3. Assign variable names to the non-reference essential nodes. This is shown in Fig. 3.2. Note that we have labeled the center node v_o, because it is the voltage we seek. Remember that these variable names represent the voltage difference between the node being labeled and the reference node.

4. Write a KCL equation at each non-reference essential node. We sum the currents leaving each node. The equations are given below. Note that since each node is a point at which three circuit components meet, each KCL equation has three terms.

$$\text{at } v_1: \qquad -5 + \frac{v_1 - v_o}{10} + 2 \quad = \quad 0$$

$$\text{at } v_o: \qquad \frac{v_o - v_1}{10} + \frac{v_o}{15} + \frac{v_o - v_2}{9} \quad = \quad 0$$

$$\text{at } v_2: \qquad \frac{v_2 - v_o}{9} + \frac{v_2}{7} - 2 \quad = \quad 0$$

Note that we have three unknowns, the three node voltages v_o, v_1, and v_2, and three equations in terms of those unknowns.

5. Write any supplemental equations. In this example, there are no supplemental equations, since there are no dependent sources or voltage sources in the circuit. Also, we already have a sufficient number of equations to solve for all of the unknowns.

6. Express the equations in standard form. The form we use collects all of the terms involving each of the unknowns on the left-hand side of each equation, and collects the constants on the right-hand side of each equation. This is shown below:

$$\text{at } v_1: \quad v_1\left(\frac{1}{10}\right) \quad + \quad v_o\left(-\frac{1}{10}\right) \quad + \quad v_2\left(0\right) \quad = \quad 3$$

$$\text{at } v_o: \quad v_1\left(-\frac{1}{10}\right) \quad + \quad v_o\left(\frac{1}{10}+\frac{1}{15}+\frac{1}{9}\right) \quad + \quad v_2\left(-\frac{1}{9}\right) \quad = \quad 0$$

$$\text{at } v_2: \quad v_1\left(0\right) \quad + \quad v_o\left(-\frac{1}{9}\right) \quad + \quad v_2\left(\frac{1}{9}+\frac{1}{7}\right) \quad = \quad 2$$

Note that there are three terms on the left-hand side of each equation, one for each of the three unknown variables. Be sure to check these equations against the KCL equations from the previous step to be certain you know how to use the standard form.

7. Solve the equations and check your solution. When these equations are input into a calculator, the solution is

$$v_1 = 60 \text{ V}; \qquad v_o = 30 \text{ V}; \qquad v_2 = 21 \text{ V}.$$

The circuit is repeated in Fig. 3.3 with the values of the node voltages labeled, and the currents through each of the branches labeled. Remember that we can calculate the current through each resistor using Ohm's law.

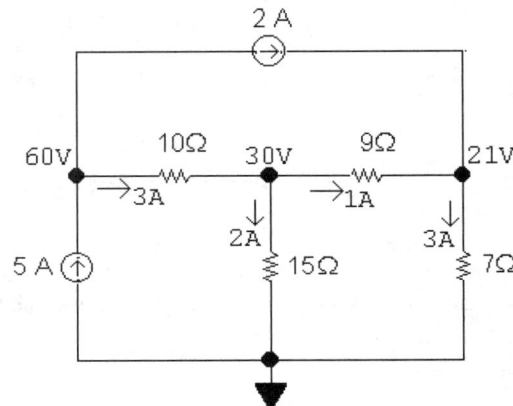

Figure 3.3: The circuit for Example 3.1, solved.

Using the values in Fig. 3.3, we can calculate the power for each component:

$$
\begin{array}{rclcl}
p_{5A} & = & -vi & = & -(60)(5) & = & -300 \text{ W}; \\
p_{2A} & = & vi & = & (60-21)(2) & = & 78 \text{ W}; \\
p_{10\Omega} & = & v^2/R & = & (60-30)^2/10 & = & 90 \text{ W}; \\
p_{15\Omega} & = & v^2/R & = & (30)^2/15 & = & 60 \text{ W}; \\
p_{9\Omega} & = & v^2/R & = & (30-21)^2/9 & = & 9 \text{ W}; \\
p_{7\Omega} & = & v^2/R & = & (21)^2/7 & = & 63 \text{ W};
\end{array}
$$

Thus,

$$\sum p = -300 + 78 + 90 + 60 + 9 + 63 = 0 \text{ W} \qquad \text{checks}$$

The power balance verifies that we have the correct solution, so $v_o = 30$ V.

Now try using the node voltage method for each of the practice problems below.

Practice Problem 3.1

Find v_o for the circuit in Fig. 3.4.

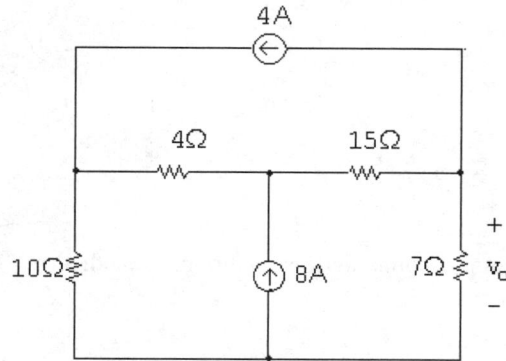

Figure 3.4: The circuit for Practice Problem 3.1.

1. Identify the essential nodes by adding black dots to Fig. 3.4.

2. Choose a reference node by adding the ground symbol to Fig. 3.4.

3. Assign variable names to the non-reference essential nodes in Fig. 3.4.

4. Write a KCL equation at each non-reference essential node.

5. Are any supplemental equations required? If not, why not? If so, write them in the space below.

6. Express all of the equations in standard form.

7. Solve the equations, using a calculator, a computer tool, or Cramer's method.

    Check your solution by calculating the power for each element and summing the power for all elements.

    Calculate v_o.

Practice Problem 3.2

Find i_o for the circuit in Fig. 3.5.

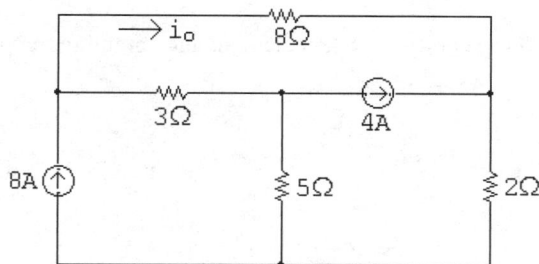

Figure 3.5: The circuit for Practice Problem 3.2.

1. Identify the essential nodes by adding black dots to Fig. 3.5.

2. Choose a reference node by adding the ground symbol to Fig. 3.5.

3. Assign variable names to the non-reference essential nodes in Fig. 3.5.

4. Write a KCL equation at each non-reference essential node.

5. Are any supplemental equations required? If not, why not? If so, write them in the space below.

6. Express all of the equations in standard form.

7. Solve the equations, using a calculator, a computer tool, or Cramer's method.

Check your solution by calculating the power for each element and summing the power for all elements.

Calculate i_o.

Practice Problem 3.3

Find v_o for the circuit in Fig. 3.6.

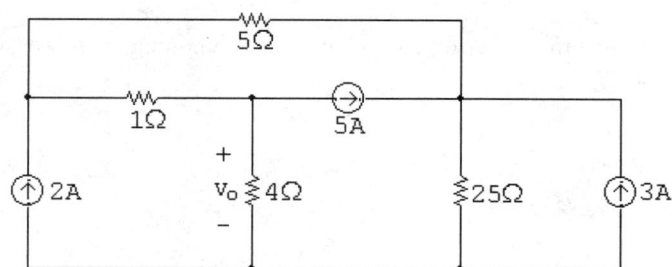

Figure 3.6: The circuit for Practice Problem 3.3.

1. Identify the essential nodes by adding black dots to Fig. 3.6.

2. Choose a reference node by adding the ground symbol to Fig. 3.6.

3. Assign variable names to the non-reference essential nodes in Fig. 3.6.

4. Write a KCL equation at each non-reference essential node.

5. Are any supplemental equations required? If not, why not? If so, write them in the space below.

6. Express all of the equations in standard form.

7. Solve the equations, using a calculator, a computer tool, or Cramer's method.

Check your solution by calculating the power for each element and summing the power for all elements.

Calculate v_o.

Practice Problem 3.4

Find i_o for the circuit in Fig. 3.7.

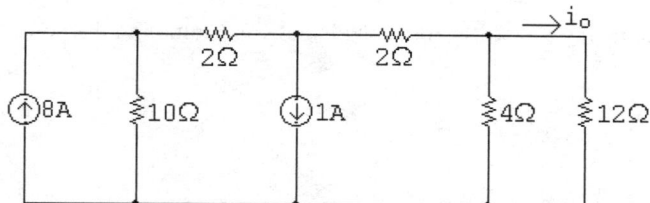

Figure 3.7: The circuit for Practice Problem 3.4.

1. Identify the essential nodes by adding black dots to Fig. 3.7.

2. Choose a reference node by adding the ground symbol to Fig. 3.7.

3. Assign variable names to the non-reference essential nodes in Fig. 3.7.

4. Write a KCL equation at each non-reference essential node.

5. Are any supplemental equations required? If not, why not? If so, write them in the space below.

6. Express all of the equations in standard form.

7. Solve the equations, using a calculator, a computer tool, or Cramer's method.

Check your solution by calculating the power for each element and summing the power for all elements.

Calculate i_o.

Practice Problem 3.5

Find the power dissipated by the 10 Ω resistor for the circuit in Fig. 3.8.

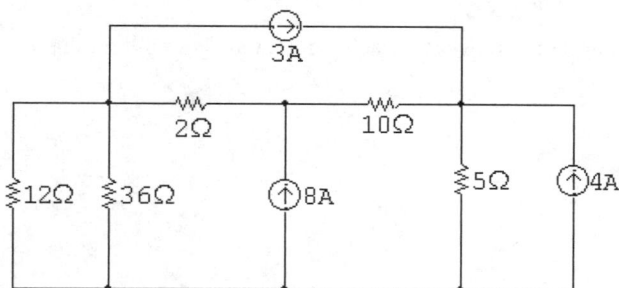

Figure 3.8: The circuit for Practice Problem 3.5.

1. Identify the essential nodes by adding black dots to Fig. 3.8.

2. Choose a reference node by adding the ground symbol to Fig. 3.8.

3. Assign variable names to the non-reference essential nodes in Fig. 3.8.

4. Write a KCL equation at each non-reference essential node.

5. Are any supplemental equations required? If not, why not? If so, write them in the space below.

6. Express all of the equations in standard form.

7. Solve the equations, using a calculator, a computer tool, or Cramer's method.

Check your solution by calculating the power for each element and summing the power for all elements.

Calculate $p_{10\Omega}$.

Example 3.2

Using the node voltage method, find v_o for the circuit in Fig. 3.9

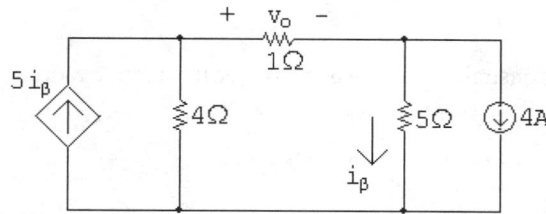

Figure 3.9: The circuit for Example 3.2

Solution

1. Identify the essential nodes. There are three points at which three or more circuit elements connect, so there are three essential nodes. They have been labeled with large black dots in Fig. 3.10.

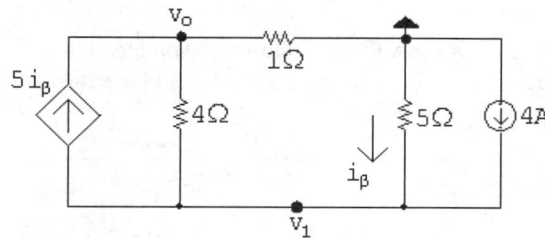

Figure 3.10: The circuit in Fig. 3.9 with the essential nodes marked, the reference node chosen, and the remaining essential nodes labeled.

2. Choose a reference node. In the circuit in Fig. 3.10 we have chosen the top right node as the reference node, and labeled it with the symbol for circuit ground. We choose this node because it is one of the two nodes associated with v_o, the voltage of interest.

3. Assign variable names to the non-reference essential nodes. This is shown in Fig. 3.10. Note that we have labeled the top left node v_o, because it is the voltage we seek. Remember that these variable names represent the voltage difference between the node being labeled and the reference node.

4. Write a KCL equation at each non-reference essential node. We sum the currents leaving each node. The equations are given below.

$$\text{at } v_o: \quad -5i_\beta + \frac{v_o - v_1}{4} + \frac{v_o}{1} = 0$$

$$\text{at } v_1: \quad 5i_\beta + \frac{v_1 - v_o}{4} + \frac{v_1}{5} - 4 = 0$$

Note that we have three unknowns, the two node voltages v_o and v_1, and the current i_β that controls the dependent source. Yet we only have two KCL equations. This means we have to specify a third equation.

5. Write any supplemental equations. This is where the third equation will be developed. Whenever there are dependent sources in our circuit, we will need to write a supplemental equation that defines the voltage or current used to control the dependent source in terms of the node voltages in our circuit. This supplemental equation is also called a **constraint equation**, because it constrains the relationship between two or more unknowns in our circuit, so that one of the unknowns is no longer an independent variable but rather is dependent on the other independent variables in our circuit.

Notice that the controlling current i_β is the current through the 5 Ω resistor, so we use Ohm's law to define this current in terms of the voltage difference across the resistor and the resistance. The constraint equation is thus

$$i_\beta = \frac{0 - v_1}{5}$$

The two KCL equations and this constraint equation now provide the three equations needed to solve for the three unknowns in the circuit.

6. Express the equations in standard form. This is shown below:

at v_o: $i_\beta(-5) \quad + \quad v_1\left(-\dfrac{1}{4}\right) \quad + \quad v_o\left(\dfrac{1}{4}+1\right) \quad = \quad 0$

at v_1: $i_\beta(5) \quad + \quad v_1\left(\dfrac{1}{4}+\dfrac{1}{5}\right) \quad + \quad v_o\left(-\dfrac{1}{4}\right) \quad = \quad 4$

constraint: $i_\beta(1) \quad + \quad v_1\left(\dfrac{1}{5}\right) \quad + \quad v_o(0) \quad = \quad 0$

7. Solve the equations and check your solution. When these equations are input into a calculator, the solution is

$$i_\beta = 2 \text{ A}; \qquad v_1 = -10 \text{ V}; \qquad v_o = 6 \text{ V}.$$

The circuit is repeated in Fig. 3.11 with the values of the node voltages labeled, and the currents through each of the branches labeled. Remember that we can calculate the current through each resistor using Ohm's law.

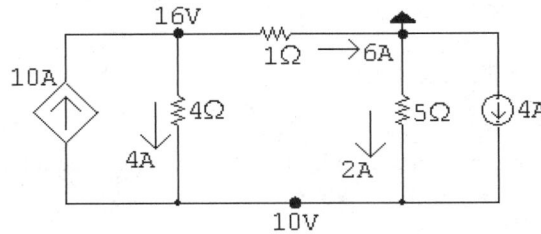

Figure 3.11: The circuit for Example 3.2, solved.

Using the values in Fig. 3.11, we can calculate the power for each component:

$$
\begin{aligned}
p_{5i_\beta} &= vi &&= (-10 - 6)[5(2)] &&= -160 \text{ W};\\
p_{4A} &= vi &&= [0 - (-10)](4) &&= 40 \text{ W};\\
p_{4\Omega} &= v^2/R &&= [6 - (-10)]^2/4 &&= 64 \text{ W};\\
p_{1\Omega} &= v^2/R &&= (6 - 0)^2/1 &&= 36 \text{ W};\\
p_{5\Omega} &= v^2/R &&= (-10 - 0)^2/5 &&= 20 \text{ W};
\end{aligned}
$$

Thus,

$$\sum p = -160 + 40 + 64 + 36 + 20 = 0 \text{ W} \qquad \text{checks}$$

The power balance verifies that we have the correct solution, so $v_o = 6$ V.

Now try using the node voltage method as it applies to circuits with dependent sources for each of the practice problems below.

Practice Problem 3.6

Find i_o for the circuit in Fig. 3.12.

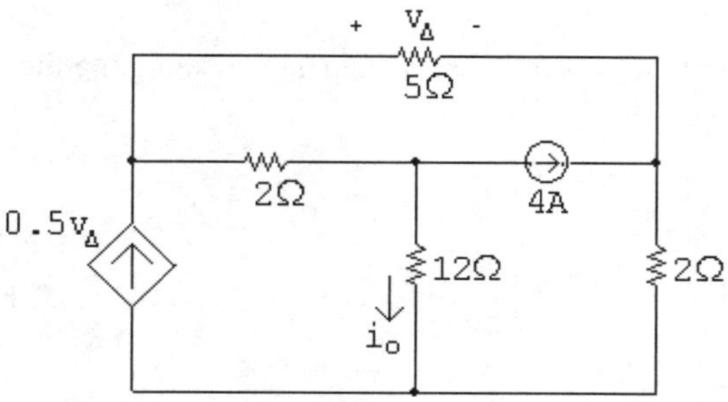

Figure 3.12: The circuit for Practice Problem 3.6.

1. Identify the essential nodes by adding black dots to Fig. 3.12.

2. Choose a reference node by adding the ground symbol to Fig. 3.12.

3. Assign variable names to the non-reference essential nodes in Fig. 3.12.

4. Write a KCL equation at each non-reference essential node.

5. Are any supplemental equations required? If not, why not? If so, write them in the space below.

6. Express all of the equations in standard form.

7. Solve the equations, using a calculator, a computer tool, or Cramer's method.

Check your solution by calculating the power for each element and summing the power for all elements.

Calculate i_o.

Practice Problem 3.7

Find v_o for the circuit in Fig. 3.13.

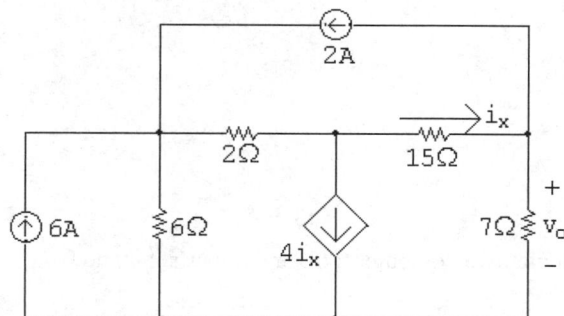

Figure 3.13: The circuit for Practice Problem 3.7.

1. Identify the essential nodes by adding black dots to Fig. 3.13.

2. Choose a reference node by adding the ground symbol to Fig. 3.13.

3. Assign variable names to the non-reference essential nodes in Fig. 3.13.

4. Write a KCL equation at each non-reference essential node.

5. Are any supplemental equations required? If not, why not? If so, write them in the space below.

6. Express all of the equations in standard form.

7. Solve the equations, using a calculator, a computer tool, or Cramer's method.

Check your solution by calculating the power for each element and summing the power for all elements.

Calculate v_o.

Practice Problem 3.8

Find the power delivered to the circuit in Fig. 3.14.

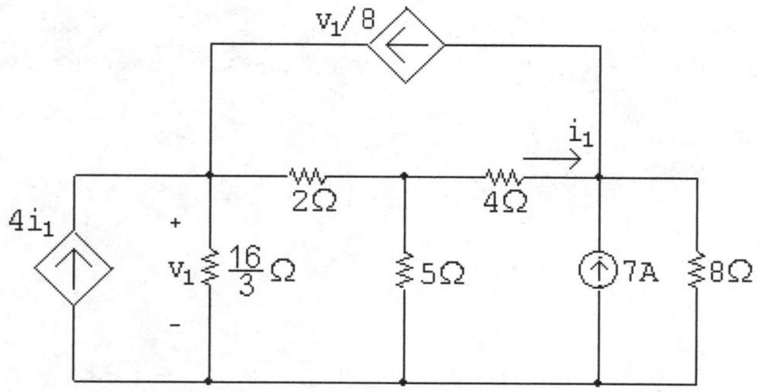

Figure 3.14: The circuit for Practice Problem 3.8.

1. Identify the essential nodes by adding black dots to Fig. 3.14.

2. Choose a reference node by adding the ground symbol to Fig. 3.14.

3. Assign variable names to the non-reference essential nodes in Fig. 3.14.

4. Write a KCL equation at each non-reference essential node.

5. Are any supplemental equations required? If not, why not? If so, write them in the space below.

6. Express all of the equations in standard form.

7. Solve the equations, using a calculator, a computer tool, or Cramer's method.

   Check your solution by calculating the power for each element and summing the power for all elements.

   Calculate the power delivered to the circuit.

Practice Problem 3.9

Find i_o for the circuit in Fig. 3.15.

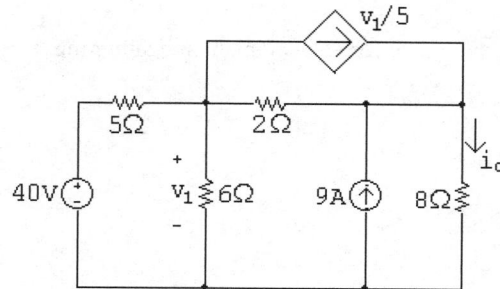

Figure 3.15: The circuit for Practice Problem 3.9.

1. Identify the essential nodes by adding black dots to Fig. 3.15.

2. Choose a reference node by adding the ground symbol to Fig. 3.15.

3. Assign variable names to the non-reference essential nodes in Fig. 3.15.

4. Write a KCL equation at each non-reference essential node.

5. Are any supplemental equations required? If not, why not? If so, write them in the space below.

6. Express all of the equations in standard form.

7. Solve the equations, using a calculator, a computer tool, or Cramer's method.

Check your solution by calculating the power for each element and summing the power for all elements.

Calculate i_o.

Practice Problem 3.10

Find v_o for the circuit in Fig. 3.16.

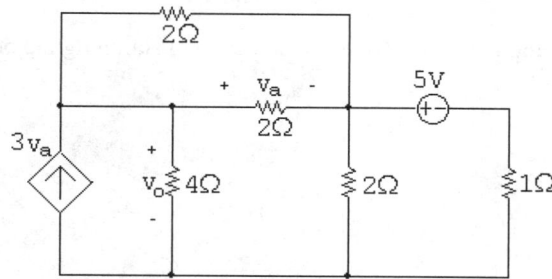

Figure 3.16: The circuit for Practice Problem 3.10.

1. Identify the essential nodes by adding black dots to Fig. 3.16.

2. Choose a reference node by adding the ground symbol to Fig. 3.16.

3. Assign variable names to the non-reference essential nodes in Fig. 3.16.

4. Write a KCL equation at each non-reference essential node. (*Hint* — interchange the positions of the 5 V source and the 2 Ω resistor.)

5. Are any supplemental equations required? If not, why not? If so, write them in the space below.

6. Express all of the equations in standard form.

7. Solve the equations, using a calculator, a computer tool, or Cramer's method.

   Check your solution by calculating the power for each element and summing the power for all elements.

   Calculate v_o.

Example 3.3

Using the node voltage method, find v_o for the circuit in Fig. 3.17

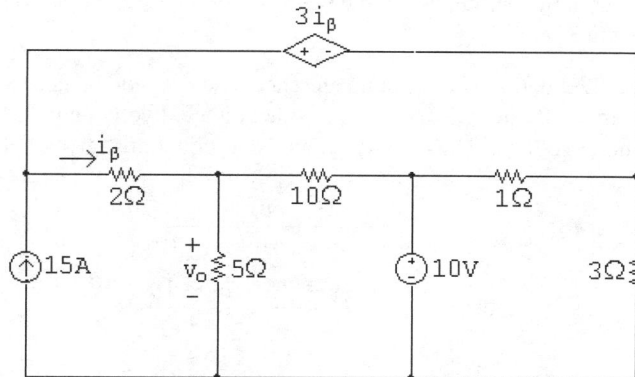

Figure 3.17: The circuit for Example 3.3

Solution

1. Identify the essential nodes. There are five points at which three or more circuit elements connect, so there are five essential nodes. They have been labeled with large black dots in Fig. 3.18.

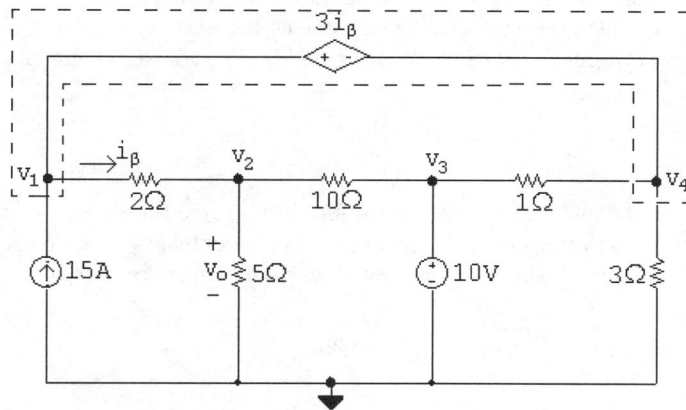

Figure 3.18: The circuit in Fig. 3.17 with the essential nodes marked, the reference node chosen, the remaining essential nodes labeled, and the supernode identified.

2. Chose a reference node. In the circuit in Fig. 3.18 we have chosen the top right node as the reference node, and labeled it with the symbol for circuit ground. We chose this node because it is one of the two nodes associated with v_o, the voltage of interest.

3. Assign variable names to the non-reference essential nodes. This is shown in Fig. 3.18. Note that the node labeled v_2 could also have been labeled v_o, as this is the node that defines the desired output with respect to the reference node.

4. Write a KCL equation at each non-reference essential node. We modify this step whenever the circuit has a voltage source between two essential nodes. This circuit has two such voltage sources. Consider first the 10 V source. Since this voltage source is between a non-reference essential node (the node labeled v_3) and the reference node, it establishes a voltage of 10 V at the non-reference essential node. Thus, $v_3 = 10$ V, so there is no need to write a KCL equation at the node labeled v_3.

Now consider the dependent voltage source. It, too, is between two essential nodes, but now neither node is the reference node. Any time a voltage source is between two non-reference essential nodes, it constrains the difference between the two voltages and forms a **supernode**. To deal with the supernode, we write one KCL equation for the supernode, and one constraint equation defining the relationship between the two node voltages that comprise the supernode. The supernode is identified by the dashed area in Fig. 3.18.

Thus, in this step we write a KCL equation at each non-reference essential node whose voltage is not known, and at each supernode. For the circuit in Fig. 3.18, there is one known node voltage, two node voltages that comprise the supernode, and one remaining unknown node voltage. Thus we write two KCL equations, given below.

$$\text{at } v_2: \qquad \frac{v_2 - v_1}{2} + \frac{v_2}{5} + \frac{v_2 - 10}{10} \quad = \quad 0$$

$$\text{at supernode:} \quad -15 + \frac{v_1 - v_2}{2} + \frac{v_4 - 10}{1} + \frac{v_4}{3} \quad = \quad 0$$

5. Write any supplemental equations. Since there is a dependent source in the circuit, we know we will need at least one supplemental equation. This equation defines the quantity used to control the dependent source, i_β in terms of the labeled node voltages. Thus, the equation is

$$i_\beta \quad = \quad \frac{v_1 - v_2}{2}$$

But we are not finished yet! Remember that the existence of a supernode means that two of the essential nodes are constrained by the voltage source in between these nodes. Therefore, every time we define a supernode in a circuit, we should expect to write a supplemental, or constraint, equation that relates the two essential nodes contained by the supernode. In this circuit, the supernode contains the essential nodes labeled v_1 and v_4, and the constraint equation defines the limitation on the voltage difference between these two nodes. The constraint equation is

$$3i_\beta = v_1 - v_4$$

The two KCL equations and the two supplemental equations provide the four independant equations needed to solve for our three unknown essential node voltages (v_1, v_2, and v_4 — remember that $v_3 = 10$ V because of the independent voltage source between v_3 and the reference node) and our unknown controlling current (i_β).

6. Express the equations in standard form. This is shown below:

$$\text{at } v_2: \qquad v_1 \left(-\frac{1}{2} \right) \quad + \quad v_2 \left(\frac{1}{2} + \frac{1}{5} + \frac{1}{10} \right) \quad + \quad v_4 \left(0 \right) \qquad + \quad i_\beta \left(0 \right) \quad = \quad 1$$

$$\text{at supernode:} \quad v_1 \left(\frac{1}{2} \right) \quad + \quad v_2 \left(-\frac{1}{2} \right) \qquad + \quad v_4 \left(1 + \frac{1}{3} \right) \quad + \quad i_\beta \left(0 \right) \quad = \quad 25$$

$$\text{dep. source:} \quad v_1 \left(-\frac{1}{2} \right) \quad + \quad v_2 \left(\frac{1}{2} \right) \qquad + \quad v_4 \left(0 \right) \qquad + \quad i_\beta \left(1 \right) \quad = \quad 0$$

$$\text{supernode:} \quad v_1 \left(-1 \right) \quad + \quad v_2 \left(0 \right) \qquad + \quad v_4 \left(1 \right) \qquad + \quad i_\beta \left(3 \right) \quad = \quad 0$$

7. Solve the equations and check your solution. When these equations are input into a calculator, the solution is

$$v_1 = 30 \text{ V}; \qquad v_2 = 20 \text{ V}; \qquad v_4 = 15 \text{ V}; \qquad i_\beta = 5\text{A}.$$

The circuit is repeated in Fig. 3.19 with the values of the node voltages labeled, and the currents through each of the branches labeled. Remember that we can calculate the current through each resistor using Ohm's law.

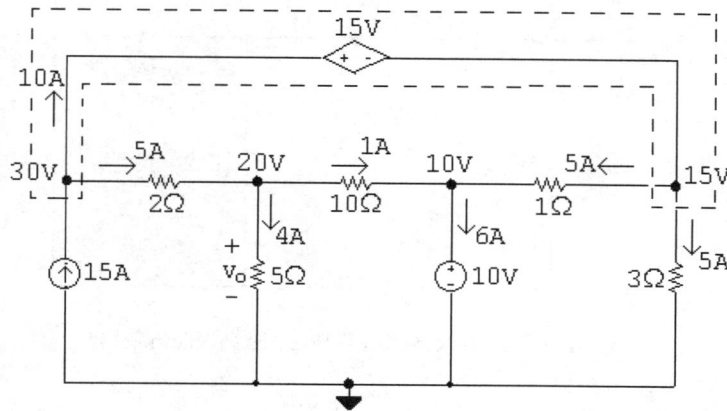

Figure 3.19: The circuit for Example 3.3, solved.

Using the values in Fig. 3.19, we can calculate the power for each component:

$$
\begin{aligned}
p_{15A} &= -vi &&= -(15)(30) &&= -450 \text{ W;} \\
p_{\text{d.s.}} &= vi &&= [3(5)](10) &&= 150 \text{ W;} \\
p_{10V} &= vi &&= (6)(10) &&= 60 \text{ W;} \\
p_{2\Omega} &= i^2 R &&= (5)^2(2) &&= 50 \text{ W;} \\
p_{5\Omega} &= i^2 R &&= (4)^2(5) &&= 80 \text{ W;} \\
p_{10\Omega} &= i^2 R &&= (1)^2(10) &&= 10 \text{ W;} \\
p_{1\Omega} &= i^2 R &&= (5)^2(1) &&= 25 \text{ W;} \\
p_{3\Omega} &= i^2 R &&= (5)^2(3) &&= 75 \text{ W;}
\end{aligned}
$$

Thus,

$$\sum p = -450 + 150 + 60 + 50 + 80 + 10 + 25 + 75 = 0 \text{ W} \qquad \text{checks}$$

The power balance verifies that we have the correct solution, so $v_o = 20$ V.

Now try using the node voltage method as it applies to circuits with voltage sources between essential nodes for each of the practice problems below.

Practice Problem 3.11

Find i_o for the circuit in Fig. 3.20.

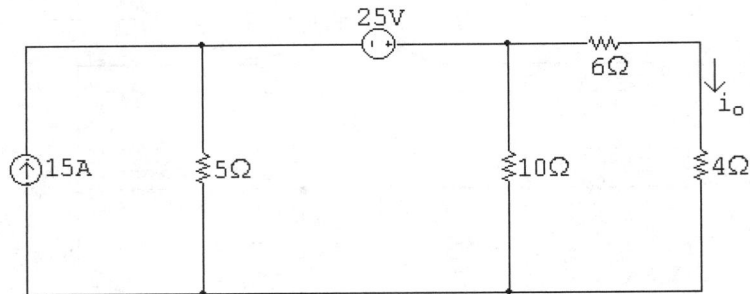

Figure 3.20: The circuit for Practice Problem 3.11.

1. Identify the essential nodes by adding black dots to Fig. 3.20.

2. Choose a reference node by adding the ground symbol to Fig. 3.20.

3. Assign variable names to the non-reference essential nodes in Fig. 3.20.

4. Write a KCL equation at each non-reference essential node for which the voltage is not already known, and at each supernode.

5. Are any supplemental equations required? If not, why not? If so, write them in the space below.

6. Express all of the equations in standard form.

7. Solve the equations, using a calculator, a computer tool, or Cramer's method.

Check your solution by calculating the power for each element and summing the power for all elements.

Calculate i_o.

Practice Problem 3.12

Find v_o for the circuit in Fig. 3.21.

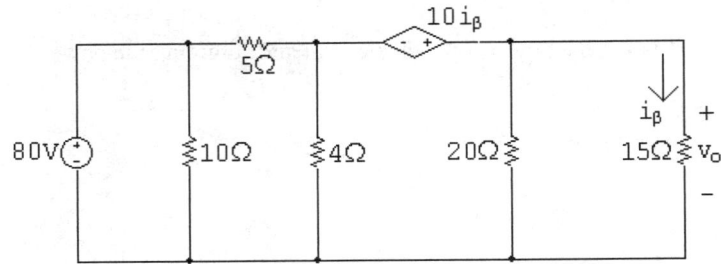

Figure 3.21: The circuit for Practice Problem 3.12.

1. Identify the essential nodes by adding black dots to Fig. 3.21.

2. Choose a reference node by adding the ground symbol to Fig. 3.21.

3. Assign variable names to the non-reference essential nodes in Fig. 3.21.

4. Write a KCL equation at each non-reference essential node for which the voltage is not already known, and at each supernode.

5. Are any supplemental equations required? If not, why not? If so, write them in the space below.

6. Express all of the equations in standard form.

7. Solve the equations, using a calculator, a computer tool, or Cramer's method.

Check your solution by calculating the power for each element and summing the power for all elements.

Calculate v_o.

Practice Problem 3.13

Find v_o for the circuit in Fig. 3.22.

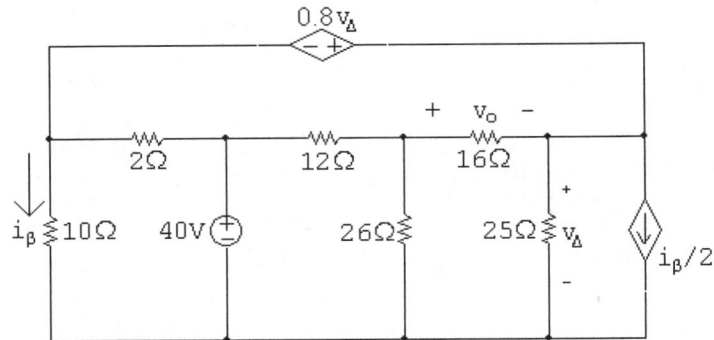

Figure 3.22: The circuit for Practice Problem 3.13.

1. Identify the essential nodes by adding black dots to Fig. 3.22.

2. Choose a reference node by adding the ground symbol to Fig. 3.22.

3. Assign variable names to the non-reference essential nodes in Fig. 3.22.

4. Write a KCL equation at each non-reference essential node for which the voltage is not already known, and at each supernode.

5. Are any supplemental equations required? If not, why not? If so, write them in the space below.

6. Express all of the equations in standard form.

7. Solve the equations, using a calculator, a computer tool, or Cramer's method.

Check your solution by calculating the power for each element and summing the power for all elements.

Calculate v_o.

Practice Problem 3.14

Find the power delivered to the circuit in Fig. 3.23.

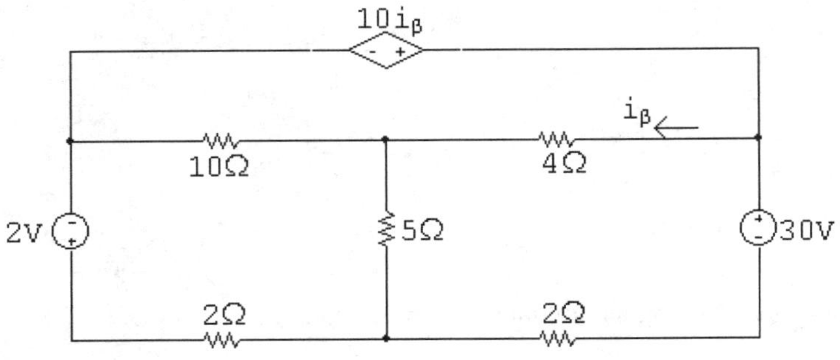

Figure 3.23: The circuit for Practice Problem 3.14.

1. Identify the essential nodes by adding black dots to Fig. 3.23.

2. Choose a reference node by adding the ground symbol to Fig. 3.23.

3. Assign variable names to the non-reference essential nodes in Fig. 3.23.

4. Write a KCL equation at each non-reference essential node for which the voltage is not already known, and at each supernode.

5. Are any supplemental equations required? If not, why not? If so, write them in the space below.

6. Express all of the equations in standard form.

7. Solve the equations, using a calculator, a computer tool, or Cramer's method.

Check your solution by calculating the power for each element and summing the power for all elements.

Calculate the power delivered to the circuit.

Practice Problem 3.15

Find v_o for the circuit in Fig. 3.24.

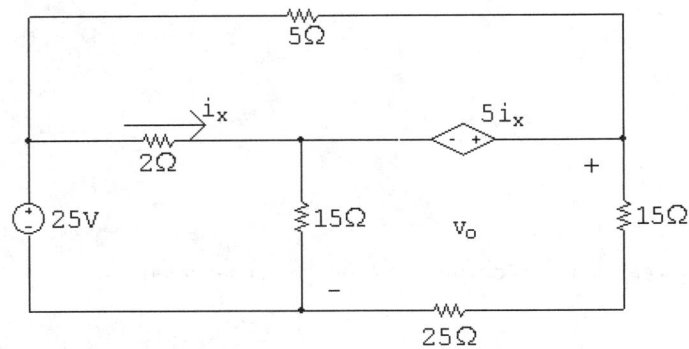

Figure 3.24: The circuit for Practice Problem 3.15.

1. Identify the essential nodes by adding black dots to Fig. 3.24.

2. Choose a reference node by adding the ground symbol to Fig. 3.24.

3. Assign variable names to the non-reference essential nodes in Fig. 3.24.

4. Write a KCL equation at each non-reference essential node for which the voltage is not already known, and at each supernode.

5. Are any supplemental equations required? If not, why not? If so, write them in the space below.

6. Express all of the equations in standard form.

7. Solve the equations, using a calculator, a computer tool, or Cramer's method.

Check your solution by calculating the power for each element and summing the power for all elements.

Calculate v_o.

Reading

- in *Electric Circuits*, ninth edition:
    ◆ Section 4.1 — terminology and definitions
    ◆ Section 4.2 — introduction to node voltage method
    ◆ Section 4.3 — node voltage method with circuits containing dependent sources
    ◆ Section 4.4 — supernodes
- Workbook section — Power Balancing in DC Circuits

Additional Problems

- 4.6 – 4.15
- 4.17 – 4.20

Solutions

- Practice Problem 3.1 — with the lower node chosen as the reference node, the node voltages are 70V, 82V, and 7V and $v_o = 7$V.

- Practice Problem 3.2 — with the lower node chosen as the reference node, the node voltages are 28V, 10V, and 12V and $i_o = 2$A.

- Practice Problem 3.3 — with the lower node chosen as the reference node, the node voltages are 20V, 12V, and 50V and $v_o = 12$V.

- Practice Problem 3.4 — with the lower node chosen as the reference node, the node voltages are 30V, 20V, and 12V and $i_o = 1$A.

- Practice Problem 3.5 — with the lower node chosen as the reference node, the node voltages are 36V, 50V, and 40V and $p_{10\Omega} = 10$W.

- Practice Problem 3.6 — with the lower node chosen as the reference node, the node voltages are 36V, 24V, and 16V and $i_o = 2$A.

- Practice Problem 3.7 — with the lower node chosen as the reference node, the node voltages are 18V, 8V, and −7V and $v_o = -7$V.

- Practice Problem 3.8 — with the lower node chosen as the reference node, the node voltages are 64V, 40V, and 24V and $p_{\text{delivered}} = 1512$W.

- Practice Problem 3.9 — with the lower node chosen as the reference node, the node voltages are 30V and 48V and $i_o = 6$A.

- Practice Problem 3.10 — with the lower node chosen as the reference node, the node voltages are 20V and 10V and $v_o = 20$V.

- Practice Problem 3.11 — with the lower node chosen as the reference node, the node voltages are 25V and 50V and $i_o = 5$A.

- Practice Problem 3.12 — with the lower node chosen as the reference node, the node voltages are 80V, 20V, and 60V and $v_o = 60$V.

- Practice Problem 3.13 — with the lower node chosen as the reference node, the node voltages are 20V, 40V, 52V, and 100V and $v_o = -48$V.

- Practice Problem 3.14 — with the lower node chosen as the reference node, the node voltages are −40V, 20V, and 60V and $p_{\text{delivered}} = 2500$W.

- Practice Problem 3.15 — with the lower node chosen as the reference node, the node voltages are 25V, 15V, and 40V and $v_o = 40$V.

Chapter 4

Mesh Current Method

The mesh current method is the companion of the node voltage method. The mesh current method, like the node voltage method, provides a systematic means to specify the equations needed to solve a circuit. Why do we have two of these systematic methods? Because for a particular circuit, one of the two methods might be easier to use, might give the desired result directly, might involve writing and solving fewer equations, or might just appeal to you more than the other method. With two methods you have a choice, and you should think through the steps of each method before deciding which one to use. Remember that once you have used one method to solve the circuit, you can use the other method to check your solution, instead of or in addition to using a power balance.

The mesh current method uses the **meshes** in a circuit. Remember that a mesh is a loop in the circuit that does not contain any other loops. The mesh current method uses **KVL** equations that are written for each of the meshes in the circuit. Remember that KVL states the the sum of all of the voltage drops around a loop is zero.

The mesh current method can be broken into the following steps:

1. Identify all of the meshes in the circuit. To do this we draw a curved arrow to identify the direction of the current flowing in the mesh.

2. Assign a variable name to the current in each mesh. Place the variable name next to the curved arrow that identifies the current and its direction. Use variable names like i_1, i_a, i_β, and so on.

3. Write a KVL equation around each of the meshes in the direction of the current arrow. We will use the same clockwise direction for each current arrow and thus will always sum the voltages in a clockwise direction.

4. Write any supplemental equations that are needed. Supplemental equations will be needed when there are dependent sources in the circuit and when there are current sources in the circuit.

5. Transform all of the equations into standard form. The standard form will enable you to solve the equations on a calculator, to solve them using a matrix method like Cramer's rule, or to solve them using a computer tool like MATLAB.

6. Solve the equations and check your solution using a power balance. If the power balances, use the solution to calculate the desired output value for the circuit.

First we present an example that contains only resistors and independent voltage sources. Once you have mastered these types of circuits we move on to circuits containing dependent voltage sources, and then to circuits containing current sources.

Example 4.1

Using the mesh current method, find i_o for the circuit in Fig. 4.1

Solution

1. Identify all of the meshes in the circuit by drawing curved arrows in the center of the mesh in the direction of the current flow. The direction of the current flow is arbitrary, but to be consistent we will always define the direction of current flow as clockwise. The current arrows are shown in Fig. 4.2.

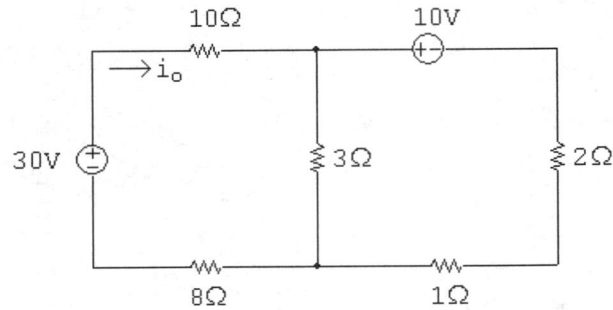

Figure 4.1: The circuit for Example 4.1

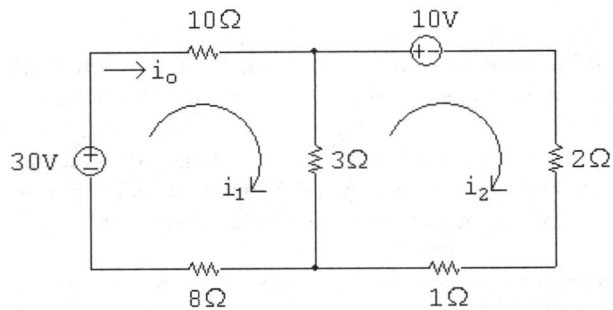

Figure 4.2: The circuit for Example 4.1, with the mesh currents defined

2. Assign a variable name for each mesh current and label the current arrow in each mesh. The chosen variable names are also shown in Fig. 4.2. Remember that the mesh currents are the currents that exist on the perimeter of each mesh. When a component belongs to only one mesh, its current is the same as the mesh current. When a component belongs to two meshes, its current is the sum of the mesh currents, where the sum must take the mesh current directions into account.

3. Write a KVL equation around each of the meshes in the direction of the current arrow. It is a good idea to put a little "x" at the point on the mesh where you start. In the left mesh, we will start just below the 30V source, and in the right mesh, we will start to the left of the 18V source.

$$\text{left mesh:} \quad -30 + 10i_1 + 3(i_1 - i_2) + 8i_2 \;=\; 0$$
$$\text{right mesh:} \quad -18 + 2i_2 + 1i_2 + 3(i_2 - i_1) \;=\; 0$$

Note that there are two unknowns, the two mesh currents i_1 and i_2, and two equations in terms of those unknowns.

4. Write any supplemental equations. In this example there are no supplemental equations, since there are no dependent sources or current sources in the circuit. Also, we have already written a sufficient number of equations to solve for all of the unknowns.

5. Place the equations in standard form. The form we use collects all of the terms involving each of the unknowns on the left-hand side of each equation, and collects the constants on the right-hand side of the equation. The standard form for the mesh current equations is shown below:

$$\text{left mesh:} \quad i_1(10 + 3 + 8) \;+\; i_2(-3) \;=\; 30$$
$$\text{right mesh:} \quad i_1(-3) \;+\; i_2(2 + 1 + 3) \;=\; 0$$

Note that there are two terms on the left-hand side of each equation, one for each of the two unknown mesh currents, and each mesh current variable appears in the same position in each equation. Be sure to check your standard form equations against your original mesh current equations to make sure you have not made an errors.

6. Solve the equations and check your solution. When these equations are input into a calculator, the solution is

$$i_1 = 2 \text{ A}; \qquad i_2 = 4 \text{ A}$$

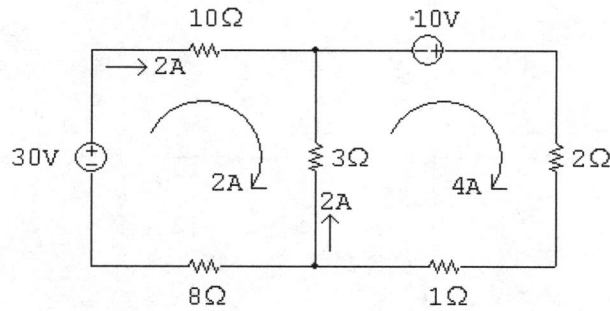

Figure 4.3: The circuit for Example 4.1, solved

The circuit is repeated in Fig. 4.3 with the values of all the currents through every component labeled. Using the values in Fig. 4.3 we can calculate the power for each component:

$$
\begin{aligned}
p_{30\text{V}} &= -vi = -(30)(2) = -60\text{ W};\\
p_{18\text{V}} &= -vi = -(18)(4) = -72\text{ W};\\
p_{10\Omega} &= i^2R = 2^2(10) = 40\text{ W};\\
p_{3\Omega} &= i^2R = 2^2(3) = 12\text{ W};\\
p_{8\Omega} &= i^2R = 2^2(8) = 32\text{ W};\\
p_{2\Omega} &= i^2R = 4^2(2) = 32\text{ W};\\
p_{1\Omega} &= i^2R = 4^2(1) = 16\text{ W};
\end{aligned}
$$

Thus,

$$
\sum p = -60 - 72 + 40 + 12 + 32 + 32 + 16 = 0\text{ W}\qquad\text{checks}
$$

The power balance verifies that we have the correct solution, so $i_o = i_1 = 2$ A.

Now try using the mesh current method for each of the practice problems below.

Practice Problem 4.1

Find i_o for the circuit in Fig. 4.4.

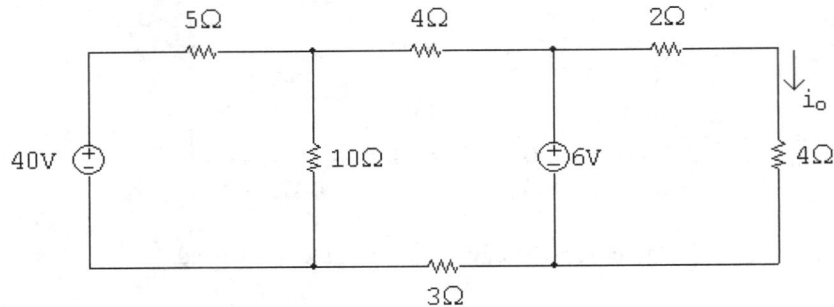

Figure 4.4: The circuit for Practice Problem 4.1.

1. Identify all of the meshes in the circuit by drawing a curved arrow in the center of each mesh in Fig. 4.4 to represent the direction of the current in that mesh.

2. Assign variable names to all of the mesh currents by labeling the mesh current arrows in Fig. 4.4.

3. Write a KVL equation around each of the meshes in the direction of the current arrow.

4. Are any supplemental equations required? If not, why not? If so, write them in the space below.

5. Express all of the equations in standard form.

6. Solve the equations, using a calculator, a computer tool, or Cramer's method.

Check your solution by calculating the power for each element and summing the power for all elements.

Calculate i_o.

Practice Problem 4.2

Find i_o for the circuit in Fig. 4.5.

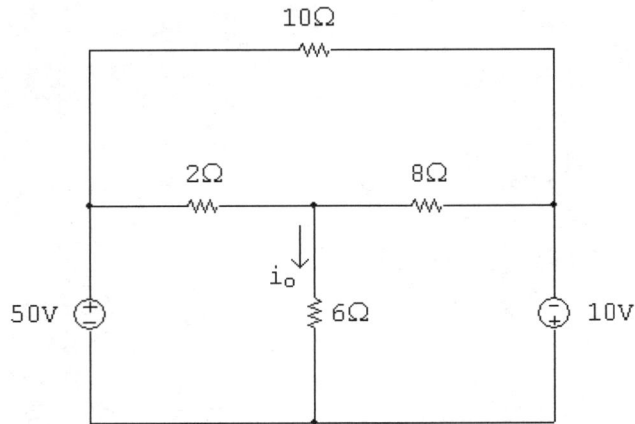

Figure 4.5: The circuit for Practice Problem 4.2.

1. Identify all of the meshes in the circuit by drawing a curved arrow in the center of each mesh in Fig. 4.5 to represent the direction of the current in that mesh.

2. Assign variable names to all of the mesh currents by labeling the mesh current arrows in Fig. 4.5.

3. Write a KVL equation around each of the meshes in the direction of the current arrow.

4. Are any supplemental equations required? If not, why not? If so, write them in the space below.

5. Express all of the equations in standard form.

6. Solve the equations, using a calculator, a computer tool, or Cramer's method.

Check your solution by calculating the power for each element and summing the power for all elements.

Calculate i_o.

Practice Problem 4.3

Find v_o for the circuit in Fig. 4.6.

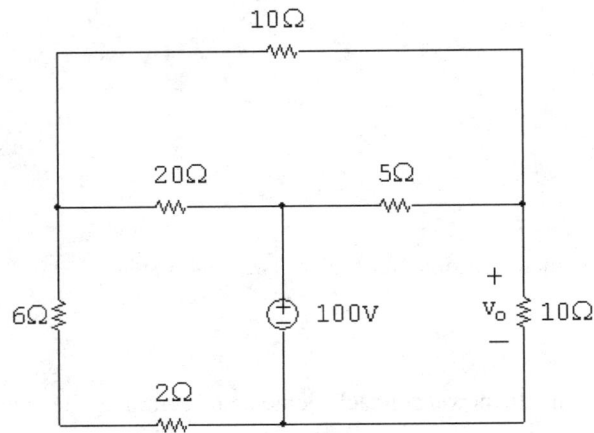

Figure 4.6: The circuit for Practice Problem 4.3.

1. Identify all of the meshes in the circuit by drawing a curved arrow in the center of each mesh in Fig. 4.6 to represent the direction of the current in that mesh.

2. Assign variable names to all of the mesh currents by labeling the mesh current arrows in Fig. 4.6.

3. Write a KVL equation around each of the meshes in the direction of the current arrow.

4. Are any supplemental equations required? If not, why not? If so, write them in the space below.

5. Express all of the equations in standard form.

6. Solve the equations, using a calculator, a computer tool, or Cramer's method.

Check your solution by calculating the power for each element and summing the power for all elements.

Calculate v_o.

Practice Problem 4.4

Find the power dissipated in the 32Ω resistor for the circuit in Fig. 4.7.

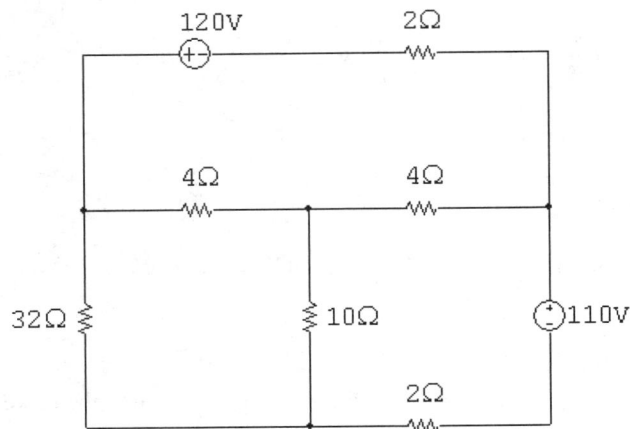

Figure 4.7: The circuit for Practice Problem 4.4.

1. Identify all of the meshes in the circuit by drawing a curved arrow in the center of each mesh in Fig. 4.7 to represent the direction of the current in that mesh.

2. Assign variable names to all of the mesh currents by labeling the mesh current arrows in Fig. 4.7.

3. Write a KVL equation around each of the meshes in the direction of the current arrow.

4. Are any supplemental equations required? If not, why not? If so, write them in the space below.

5. Express all of the equations in standard form.

6. Solve the equations, using a calculator, a computer tool, or Cramer's method.

Check your solution by calculating the power for each element and summing the power for all elements.

Calculate $p_{32\Omega}$.

Practice Problem 4.5

Find v_o for the circuit in Fig. 4.8.

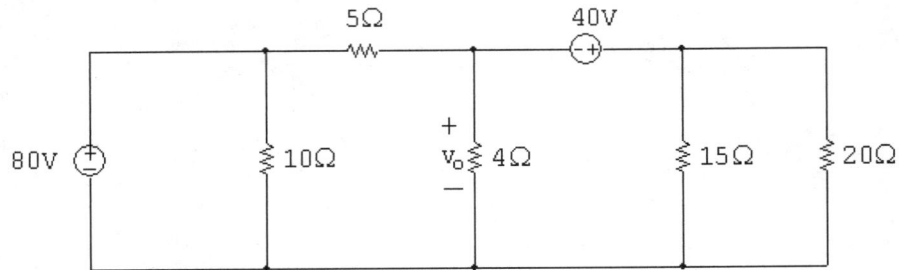

Figure 4.8: The circuit for Practice Problem 4.5.

1. Identify all of the meshes in the circuit by drawing a curved arrow in the center of each mesh in Fig. 4.8 to represent the direction of the current in that mesh.

2. Assign variable names to all of the mesh currents by labeling the mesh current arrows in Fig. 4.8.

3. Write a KVL equation around each of the meshes in the direction of the current arrow.

4. Are any supplemental equations required? If not, why not? If so, write them in the space below.

5. Express all of the equations in standard form.

6. Solve the equations, using a calculator, a computer tool, or Cramer's method.

Check your solution by calculating the power for each element and summing the power for all elements.

Calculate v_o.

Example 4.2

Using the mesh current method, find i_o for the circuit in Fig. 4.9

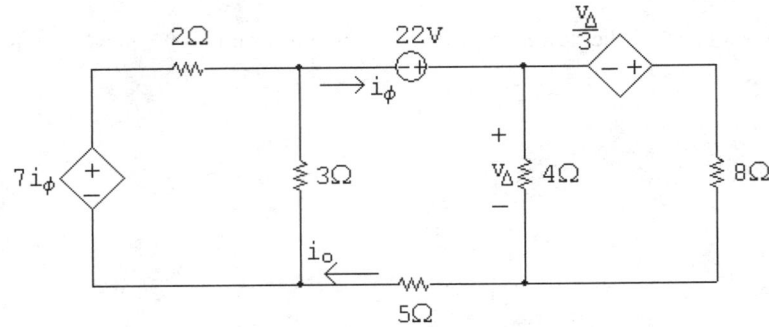

Figure 4.9: The circuit for Example 4.2

Solution

1. Identify all of the meshes in the circuit by drawing curved arrows in the center of the mesh in the direction of the current flow. As usual, we define the direction of current flow as clockwise. The current arrows are shown in Fig. 4.10.

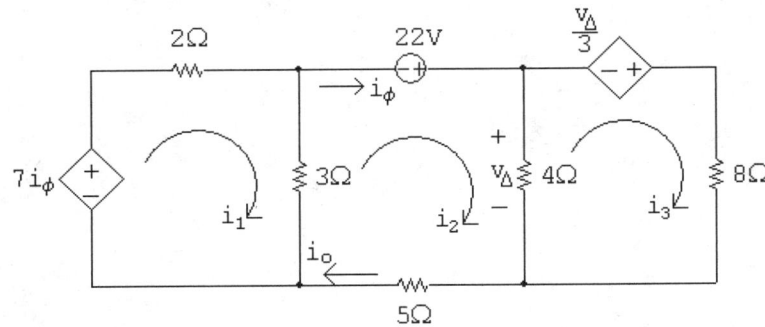

Figure 4.10: The circuit for Example 4.2, with the mesh currents defined

2. Assign a variable name for each mesh current and label the current arrow in each mesh. The chosen variable names are also shown in Fig. 4.10.

3. Write a KVL equation around each of the meshes in the direction of the current arrow. In the left mesh, we will start just below the dependent source, in the center mesh we start just to the left of the 22V source, and in the right mesh we will start to the left of the dependent source.

$$
\begin{array}{rl}
\text{left mesh:} & -7i_\phi + 2i_1 + 3(i_1 - i_2) = 0 \\
\text{center mesh:} & -22 + 4(i_2 - i_3) + 5i_2 + 3(i_2 - i_1) = 0 \\
\text{right mesh:} & -\dfrac{v_\Delta}{3} + 8i_3 + 4(i_3 - i_2) = 0
\end{array}
$$

Note that there are five unknowns, the three mesh currents i_1, i_2, and i_3, the current i_ϕ that controls one dependent source and the voltage v_Δ that controls the other dependent source. Yet there are only three KVL equations. This means we have to specify two more equations.

4. Write any supplemental equations. This is where the remaining two equations will be developed. Whenever there are dependent sources in the circuit, we will need to write a supplemental equation for each dependent source that defines the current or voltage used to control each source in terms of the mesh currents in our circuit. These supplemental equations are also called **constraint equations** because they constrain the relationship between two or more unknowns in our circuit.

Thus, one of the unknowns is no longer an independent variable byt rather is dependent on the other independent variables in the circuit.

Now turn to the circuit in Fig. 4.10. Notice that the controlling current i_ϕ is the same as the mesh current i_2. Thus, our first constraint equation is

$$i_\phi = i_2$$

From the circuit we see that the controlling voltage v_Δ is the voltage drop across the 4Ω resistor. We use Ohm's Law to define that voltage drop in terms of the current flowing through the resistor. The current flowing through the resistor in the direction of the voltage drop must be defined in terms of the mesh currents, so is equal to $i_2 - i_3$. Thus, our second constraint equation is

$$v_\Delta = 4(i_2 - i_3)$$

The three KVL equations and the two constraint equations now provide the five equations needed to solve for the five unknowns in the circuit.

5. Place the equations in standard form. This is shown below:

left mesh:	$i_1(2+3)$ +	$i_2(-3)$ +	$i_3(0)$ +	$i_\phi(-7)$ +	$v_\Delta(0)$ = 0
center mesh:	$i_1(-3)$ +	$i_2(3+4+5)$ +	$i_3(-4)$ +	$i_\phi(0)$ +	$v_\Delta(0)$ = 22
right mesh:	$i_1(0)$ +	$i_2(-4)$ +	$i_3(4+8)$ +	$i_\phi(0)$ +	$v_\Delta(-1/3)$ = 0
constraint:	$i_1(0)$ +	$i_2(1)$ +	$i_3(0)$ +	$i_\phi(-1)$ +	$v_\Delta(0)$ = 0
constraint:	$i_1(0)$ +	$i_2(-4)$ +	$i_3(4)$ +	$i_\phi(0)$ +	$v_\Delta(1)$ = 0

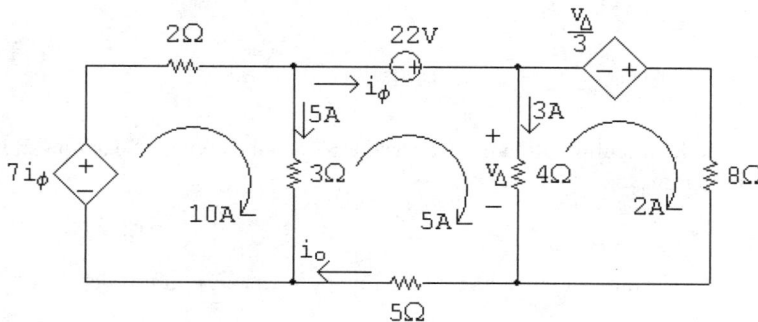

Figure 4.11: The circuit for Example 4.2, solved

6. Solve the equations and check your solution. When these equations are input into a calculator, the solution is

$$i_1 = 10 \text{ A}; \qquad i_2 = 5 \text{ A}; \qquad i_3 = 2 \text{ A}; \qquad i_\phi = 5 \text{ A}; \qquad v_\Delta = 12 \text{ V}$$

The circuit is repeated in Fig. 4.11 with the values of all the currents through every component labeled. Using the values in Fig. 4.11 we can calculate the power for each component:

$$
\begin{aligned}
p_{7i_\phi} &= -vi = -[7(5)](10) = -350 \text{ W}; \\
p_{22V} &= -vi = -(22)(5) = -110 \text{ W}; \\
p_{v_\Delta/3} &= -vi = -[12/3](2) = -8 \text{ W}; \\
p_{2\Omega} &= i^2R = 10^2(2) = 200 \text{ W}; \\
p_{3\Omega} &= i^2R = 5^2(3) = 75 \text{ W}; \\
p_{5\Omega} &= i^2R = 5^2(5) = 125 \text{ W}; \\
p_{4\Omega} &= i^2R = 3^2(4) = 36 \text{ W}; \\
p_{8\Omega} &= i^2R = 2^2(8) = 32 \text{ W};
\end{aligned}
$$

Thus,

$$\sum p = -350 - 110 - 8 + 200 + 75 + 125 + 36 + 32 = 0 \text{ W} \qquad \text{checks}$$

The power balance verifies that we have the correct solution, so $i_o = i_2 = 5$ A.

Now try using the mesh current method for each of the practice problems below.

Practice Problem 4.6

Find i_o for the circuit in Fig. 4.12.

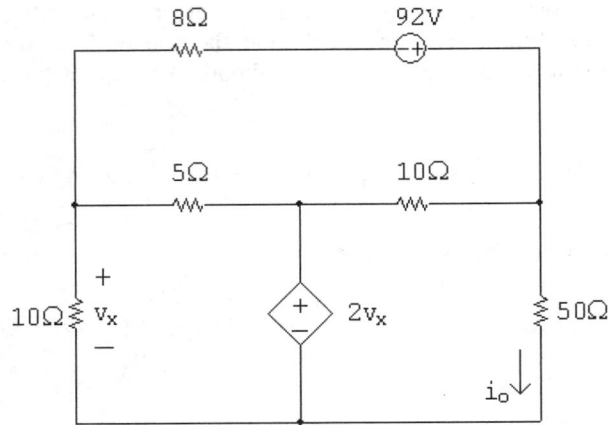

Figure 4.12: The circuit for Practice Problem 4.6.

1. Identify all of the meshes in the circuit by drawing a curved arrow in the center of each mesh in Fig. 4.12 to represent the direction of the current in that mesh.

2. Assign variable names to all of the mesh currents by labeling the mesh current arrows in Fig. 4.12.

3. Write a KVL equation around each of the meshes in the direction of the current arrow.

4. Are any supplemental equations required? If not, why not? If so, write them in the space below.

5. Express all of the equations in standard form.

6. Solve the equations, using a calculator, a computer tool, or Cramer's method.

Check your solution by calculating the power for each element and summing the power for all elements.

Calculate i_o.

Practice Problem 4.7

Find v_o for the circuit in Fig. 4.13.

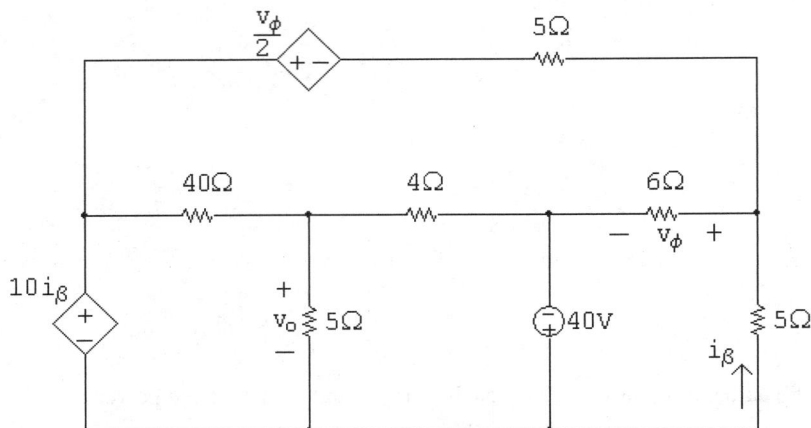

Figure 4.13: The circuit for Practice Problem 4.7.

1. Identify all of the meshes in the circuit by drawing a curved arrow in the center of each mesh in Fig. 4.13 to represent the direction of the current in that mesh.

2. Assign variable names to all of the mesh currents by labeling the mesh current arrows in Fig. 4.13.

3. Write a KVL equation around each of the meshes in the direction of the current arrow.

4. Are any supplemental equations required? If not, why not? If so, write them in the space below.

5. Express all of the equations in standard form.

6. Solve the equations, using a calculator, a computer tool, or Cramer's method.

Check your solution by calculating the power for each element and summing the power for all elements.

Calculate v_o.

Practice Problem 4.8

Find the power delivered in the circuit in Fig. 4.14.

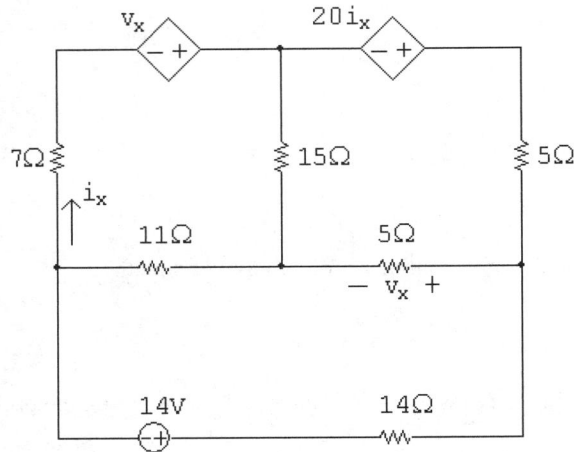

Figure 4.14: The circuit for Practice Problem 4.8.

1. Identify all of the meshes in the circuit by drawing a curved arrow in the center of each mesh in Fig. 4.14 to represent the direction of the current in that mesh.

2. Assign variable names to all of the mesh currents by labeling the mesh current arrows in Fig. 4.14.

3. Write a KVL equation around each of the meshes in the direction of the current arrow.

4. Are any supplemental equations required? If not, why not? If so, write them in the space below.

5. Express all of the equations in standard form.

6. Solve the equations, using a calculator, a computer tool, or Cramer's method.

Check your solution by calculating the power for each element and summing the power for all elements.

Calculate $p_{\text{delivered}}$.

Practice Problem 4.9

Find the power for the 15Ω resistor in the circuit in Fig. 4.15.

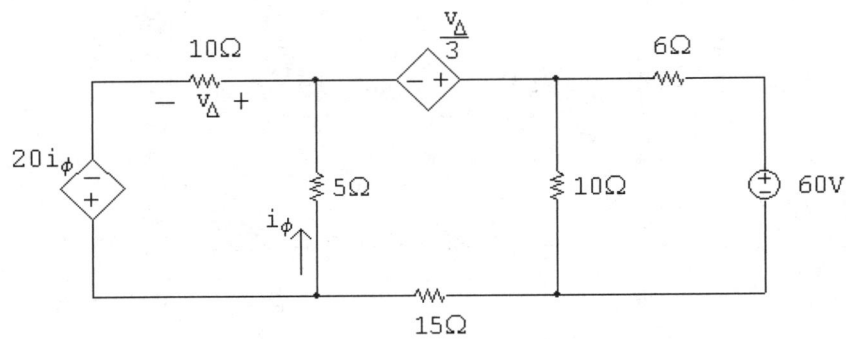

Figure 4.15: The circuit for Practice Problem 4.9.

1. Identify all of the meshes in the circuit by drawing a curved arrow in the center of each mesh in Fig. 4.15 to represent the direction of the current in that mesh.

2. Assign variable names to all of the mesh currents by labeling the mesh current arrows in Fig. 4.15.

3. Write a KVL equation around each of the meshes in the direction of the current arrow.

4. Are any supplemental equations required? If not, why not? If so, write them in the space below.

5. Express all of the equations in standard form.

6. Solve the equations, using a calculator, a computer tool, or Cramer's method.

Check your solution by calculating the power for each element and summing the power for all elements.

Calculate $p_{15\Omega}$.

Practice Problem 4.10

Find v_o in the circuit in Fig. 4.16.

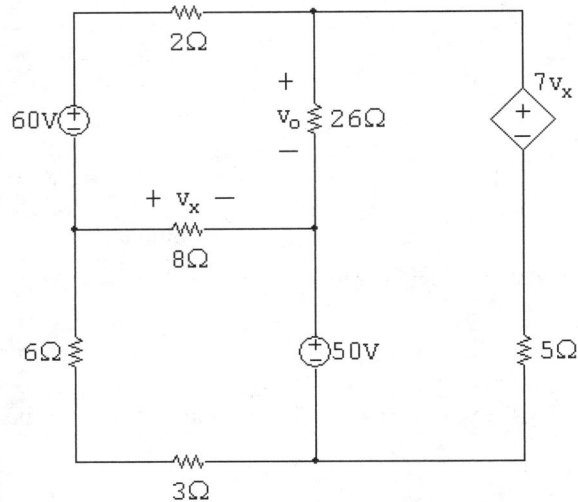

Figure 4.16: The circuit for Practice Problem 4.10.

1. Identify all of the meshes in the circuit by drawing a curved arrow in the center of each mesh in Fig. 4.16 to represent the direction of the current in that mesh.

2. Assign variable names to all of the mesh currents by labeling the mesh current arrows in Fig. 4.16.

3. Write a KVL equation around each of the meshes in the direction of the current arrow.

4. Are any supplemental equations required? If not, why not? If so, write them in the space below.

5. Express all of the equations in standard form.

6. Solve the equations, using a calculator, a computer tool, or Cramer's method.

Check your solution by calculating the power for each element and summing the power for all elements.

Calculate v_o.

Example 4.3

Using the mesh current method, find i_o for the circuit in Fig. 4.17

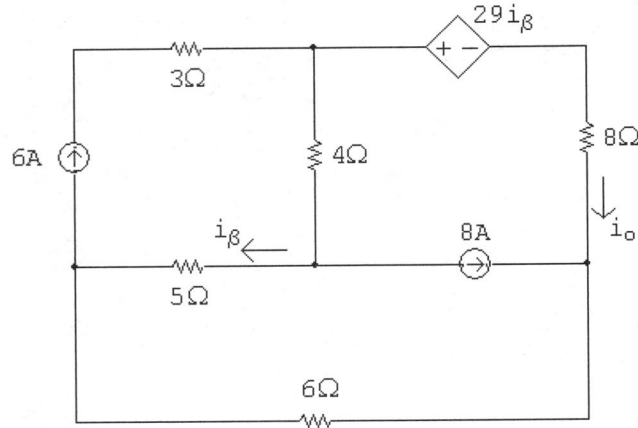

Figure 4.17: The circuit for Example 4.3

Solution

1. Identify all of the meshes in the circuit by drawing curved arrows in the center of the mesh in the direction of the current flow. As usual, we define the direction of current flow as clockwise. The current arrows are shown in Fig. 4.18.

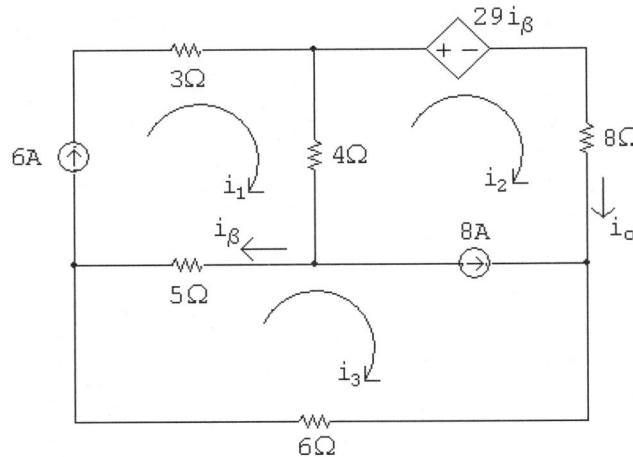

Figure 4.18: The circuit for Example 4.3, with the mesh currents defined.

2. Assign a variable name for each mesh current and label the current arrow in each mesh. The chosen variable names are also shown in Fig. 4.18.

3. Write a KVL equation around each of the meshes in the direction of the current arrow. We modify this step whenever the circuit contains current sources. The circuit in Fig. 4.18 has two current sources. Consider first the 6A current source. This current source is on the perimeter of a mesh, meaning that the current source establishes the value of the mesh current in this mesh. Thus, $i_1 = 6$A, so there is no need to write a KVL equation for this mesh.

   Now consider the 8A current source. This source is shared between two meshes, rather than being on the perimeter of a single mesh. Any time a current source is shared between two meshes, the two meshes should be combined to form a **supermesh**. Whenever a supermesh is present in a circuit we will write one single KVL equation for the supermesh and one constraint equation defining the relationship between the two mesh currents that form the supermesh. Figure 4.19 shows the known value of the current in the top left mesh and identifies the path of the supermesh with a dashed line.

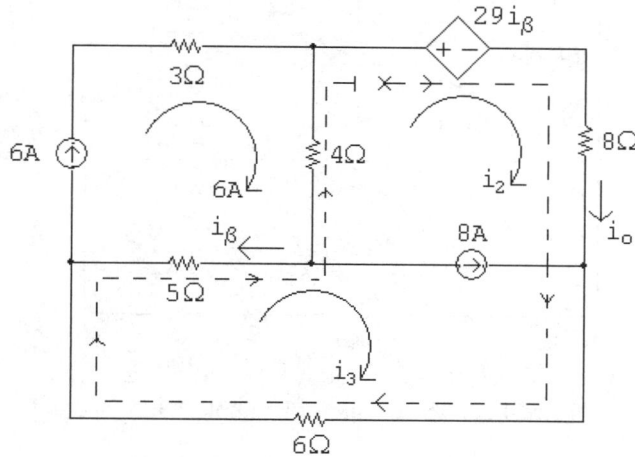

Figure 4.19: The circuit for Example 4.3, with a known mesh current and a supermesh.

Thus, in this step we write a KVL equation for each single mesh where the current is not known and for each supermesh. For the circuit in Fig. 4.19 we need only write the single KVL equation for the supermesh, because the remaining mesh current is known. We start just to the left of the dependent voltage source:

$$\text{supermesh:} \quad 29i_\beta + 8i_2 + 6i_3 + 5(i_3 - 6A) + 4(i_2 - 6A) \quad = \quad 0$$

4. Write any supplemental equations. Since there is a dependent source in the circuit, we know we will need at least one supplemental equation. This equation defines the quantity used to control the dependent source, i_β, in terms of the labeled mesh currents. Thus, the equation is

$$i_\beta = 6A - i_3$$

But there is a second supplemental, or constraint, equation due to the presence of the supermesh. Remember that the current source shared between the two meshes constrains the difference between these mesh currents. The second constraint equation is thus

$$i_3 - i_2 = 8 \text{ A}$$

The single KVL equation and the two supplemental equations provide the three equations needed to solve for the three unknowns — i_2, i_3, and i_β. Remember that $i_1 = 6A$ because of the current source on the perimeter of the top left mesh.

5. Solve the equations and check your solution. When these equations are input into a calculator, the solution is

$$i_2 = -4 \text{ A}; \qquad i_3 = 4 \text{ A}; \qquad i_\beta = 2 \text{ A}$$

The circuit is repeated in Fig. 4.20 with the values of all the currents through every component labeled. In addition, we have labeled the voltage drop across each current source. The voltage drops were calculated by writing a KVL equation for a mesh containing the current source and treating the voltage drop across the current source as an unknown. For example, to calculate the voltage drop across the 6A current source, define the voltage drop as v_{6A} (positive at the top) and write a KVL equation for the top left mesh, starting just below the 6A source and going clockwise:

$$-v_{6A} + (3\Omega)(6A) + (4\Omega)(10A) + (5\Omega)(2A) = 0$$

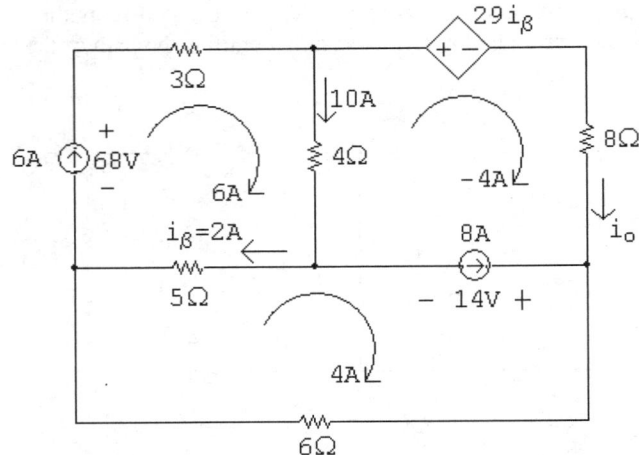

Figure 4.20: The circuit for Example 4.3, solved

Solving, we see that $v_{6A} = 68$V, as indicated in Fig. 4.20. Using the values in Fig. 4.20 we can calculate the power for each component:

$$
\begin{aligned}
p_{6A} &= -vi &=& \quad -(68)(6) &=& \quad -408 \text{ W};\\
p_{8A} &= -vi &=& \quad -(14)(8) &=& \quad -112 \text{ W};\\
p_{29i_\beta} &= vi &=& \quad [29(2)](-4) &=& \quad -232 \text{ W};\\
p_{3\Omega} &= i^2 R &=& \quad 6^2(3) &=& \quad 108 \text{ W};\\
p_{4\Omega} &= i^2 R &=& \quad 10^2(4) &=& \quad 400 \text{ W};\\
p_{5\Omega} &= i^2 R &=& \quad 2^2(5) &=& \quad 20 \text{ W};\\
p_{6\Omega} &= i^2 R &=& \quad 4^2(6) &=& \quad 96 \text{ W};\\
p_{8\Omega} &= i^2 R &=& \quad 4^2(8) &=& \quad 128 \text{ W};
\end{aligned}
$$

Thus,

$$\sum p = -408 - 112 - 232 + 108 + 400 + 20 + 96 + 128 = 0 \text{ W} \qquad \text{checks}$$

The power balance verifies that we have the correct solution, so $i_o = i_2 = -4$ A.

Now try using the mesh current method for each of the practice problems below.

Practice Problem 4.11

Find the power delivered to the 18Ω resistor in the circuit in Fig. 4.21.

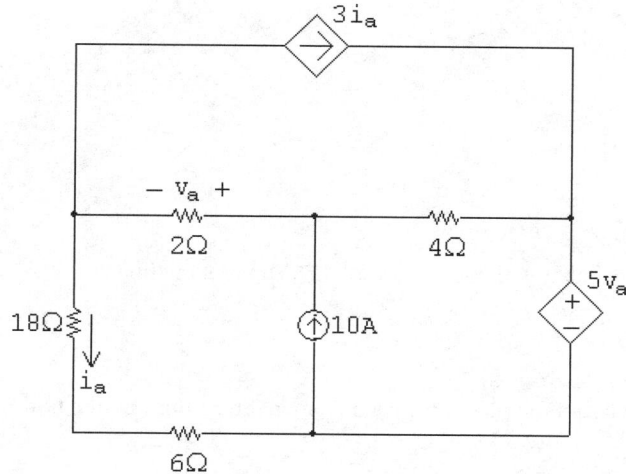

Figure 4.21: The circuit for Practice Problem 4.11.

1. Identify all of the meshes in the circuit by drawing a curved arrow in the center of each mesh in Fig. 4.21 to represent the direction of the current in that mesh.

2. Assign variable names to all of the mesh currents by labeling the mesh current arrows in Fig. 4.21.

3. Write a KVL equation around each of the meshes in the direction of the current arrow.

4. Are any supplemental equations required? If not, why not? If so, write them in the space below.

5. Express all of the equations in standard form.

6. Solve the equations, using a calculator, a computer tool, or Cramer's method.

Check your solution by calculating the power for each element and summing the power for all elements.

Calculate $p_{18\Omega}$.

Practice Problem 4.12

Find i_o for the circuit in Fig. 4.22.

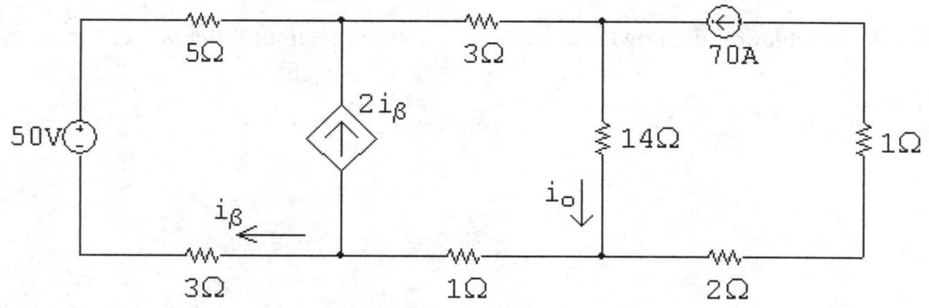

Figure 4.22: The circuit for Practice Problem 4.12.

1. Identify all of the meshes in the circuit by drawing a curved arrow in the center of each mesh in Fig. 4.22 to represent the direction of the current in that mesh.

2. Assign variable names to all of the mesh currents by labeling the mesh current arrows in Fig. 4.22.

3. Write a KVL equation around each of the meshes in the direction of the current arrow.

4. Are any supplemental equations required? If not, why not? If so, write them in the space below.

5. Express all of the equations in standard form.

6. Solve the equations, using a calculator, a computer tool, or Cramer's method.

Check your solution by calculating the power for each element and summing the power for all elements.

Calculate i_o.

Practice Problem 4.13

Find v_o for the circuit in Fig. 4.23.

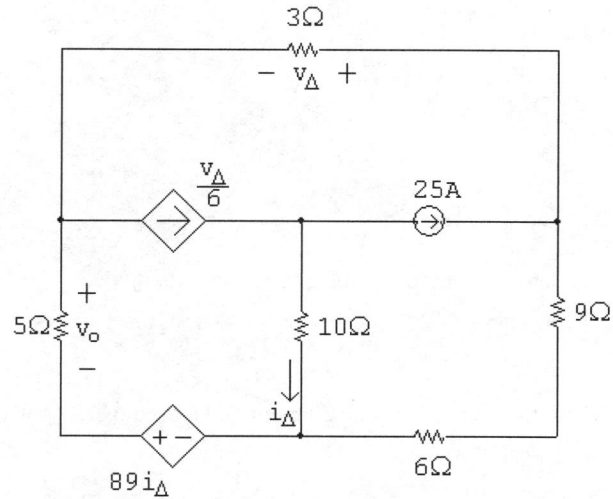

Figure 4.23: The circuit for Practice Problem 4.13.

1. Identify all of the meshes in the circuit by drawing a curved arrow in the center of each mesh in Fig. 4.23 to represent the direction of the current in that mesh.

2. Assign variable names to all of the mesh currents by labeling the mesh current arrows in Fig. 4.23.

3. Write a KVL equation around each of the meshes in the direction of the current arrow.

4. Are any supplemental equations required? If not, why not? If so, write them in the space below.

5. Express all of the equations in standard form.

6. Solve the equations, using a calculator, a computer tool, or Cramer's method.

Check your solution by calculating the power for each element and summing the power for all elements.

Calculate v_o.

Practice Problem 4.14

Find the power for the 80V source in the circuit in Fig. 4.24.

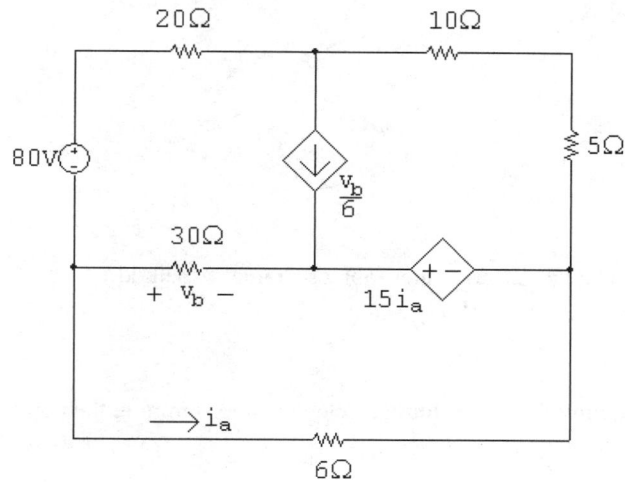

Figure 4.24: The circuit for Practice Problem 4.14.

1. Identify all of the meshes in the circuit by drawing a curved arrow in the center of each mesh in Fig. 4.24 to represent the direction of the current in that mesh.

2. Assign variable names to all of the mesh currents by labeling the mesh current arrows in Fig. 4.24.

3. Write a KVL equation around each of the meshes in the direction of the current arrow.

4. Are any supplemental equations required? If not, why not? If so, write them in the space below.

5. Express all of the equations in standard form.

6. Solve the equations, using a calculator, a computer tool, or Cramer's method.

   Check your solution by calculating the power for each element and summing the power for all elements.

   Calculate p_{80V}.

Practice Problem 4.15

Find i_o for the circuit in Fig. 4.25.

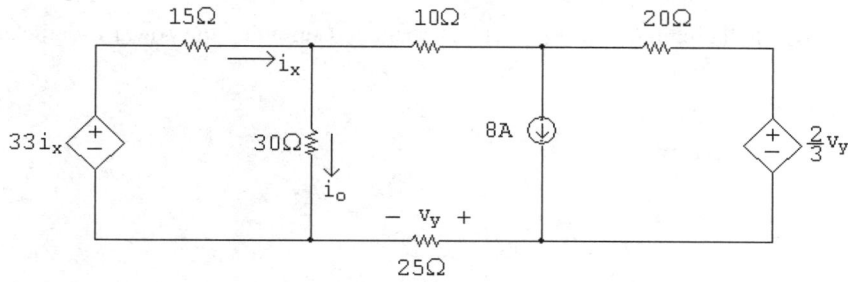

Figure 4.25: The circuit for Practice Problem 4.15.

1. Identify all of the meshes in the circuit by drawing a curved arrow in the center of each mesh in Fig. 4.25 to represent the direction of the current in that mesh.

2. Assign variable names to all of the mesh currents by labeling the mesh current arrows in Fig. 4.25.

3. Write a KVL equation around each of the meshes in the direction of the current arrow.

4. Are any supplemental equations required? If not, why not? If so, write them in the space below.

5. Express all of the equations in standard form.

6. Solve the equations, using a calculator, a computer tool, or Cramer's method.

Check your solution by calculating the power for each element and summing the power for all elements.

Calculate i_o.

Reading

- in *Electric Circuits*, ninth edition:

  ◆ Section 4.1 — terminology and definitions

  ◆ Section 4.5 — introduction to mesh current method

  ◆ Section 4.6 — mesh current method with circuits containing dependent sources

  ◆ Section 4.7 — supermeshes

- Workbook section — Power Balancing in DC Circuits

Additional Problems

- 4.31 – 4.51

Solutions

- Practice Problem 4.1 — the clockwise mesh currents are 4A, 2A, and 1A and $i_o = 1$A.

- Practice Problem 4.2 — the clockwise mesh currents are 16A, 6A, and 11A and $i_o = 5$A.

- Practice Problem 4.3 — the clockwise mesh currents are -5A, -2A, and 6A and $v_o = 60$V.

- Practice Problem 4.4 — the clockwise mesh currents are -5A, -20A, and -15A and $p_{32\Omega} = 800$W.

- Practice Problem 4.5 — the clockwise mesh currents are 20A, 12A, 7A, and 3A and $v_o = 20$V.

- Practice Problem 4.6 — the clockwise mesh currents are -4A, 4A, and 2A and $i_o = 2$A.

- Practice Problem 4.7 — the clockwise mesh currents are 4A, 8A, -2A, and 3A and $v_o = -20$V.

- Practice Problem 4.8 — the clockwise mesh currents are 4A, 6A, and 2A and $p_{\text{delivered}} = 560$W.

- Practice Problem 4.9 — the clockwise mesh currents are -6A, -2A, and -5A and $p_{15\Omega} = 60$W.

- Practice Problem 4.10 — the clockwise mesh currents are -15A, -10A and -20A and $v_o = 130$V.

- Practice Problem 4.11 — the clockwise mesh currents are -5A, 15A, and 5A and $p_{18\Omega} = 450$W.

- Practice Problem 4.12 — the clockwise mesh currents are -15A, -45A, and -70A and $i_o = 25$A.

- Practice Problem 4.13 — the clockwise mesh currents are -20A, -40A, and -15A and $v_o = 100$V.

- Practice Problem 4.14 — the clockwise mesh currents are 7A, -8A, and 10A and $p_{80\text{V}} = 560$W (delivered).

- Practice Problem 4.15 — the clockwise mesh currents are 15A, 6A, and -2A and $i_o = 9$A.

Chapter 5

Thévenin and Norton Equivalents

The **Thévenin equivalent** method allows you to replace any circuit consisting of independent sources, dependent sources, and resistors with a simple circuit consisting of a single voltage source in series with a single resistor where the simple circuit is **equivalent** to the original circuit. This means that a resistor first attached to the original circuit and then attached to the simple circuit could not distinguish between the two circuits, since the resistor would experience the same voltage drop, the same current flow, and thus the same power dissipation. The Thévenin equivalent method can thus be used to reduce the complexity of a circuit and make it much easier to analyze. A **Norton equivalent** circuit consists of a single current source in parallel with a single resistor and can be constructed from a Thévenin equivalent circuit using source transformation. Thus in this section we will present a technique for calculating the component values for a Thévenin equivalent circuit; if you want the Norton equivalent circuit, you can calculate the Thévenin equivalent circuit and use source transformation.

There are three important quantities that make up a Thévenin equivalent circuit: the open-circuit voltage, v_{oc}, the short-circuit current, i_{sc}, and the Thévenin equivalent resistance, R_{Th}. In the Thévenin equivalent circuit, the value of the voltage source is v_{oc} and the value of the series resistor is R_{Th}. In the Norton equivalent circuit, the value of the current source is i_{sc} and the value of the parallel resistor is R_{Th}. But it is not necessary to calculate all three quantities, since they are related by the following equation:

$$v_{oc} = R_{Th} i_{sc}.$$

Thus we need to determine just two of these three quantities, and can use their relationship to find the third quantity, if desired.

In circuits containing only independent sources and resistors, our Thévenin equivalent method will determine the values of v_{oc} and R_{Th}. When a circuit also contains dependent sources we will modify the method and determine v_{oc} and i_{sc}. In the examples and practice problems that follow we will calculate the Thévenin equivalent or Norton equivalent circuit as seen from a single load resistor. We will then reattach the load resistor to the Thévenin equivalent or Norton equivalent circuit and analyze this simplified circuit to determine a requested quantity.

The Thévenin equivalent method can be broken into the following steps:

1. First calculate the open-circuit voltage. Draw the circuit with the load resistor removed, which creates an open circuit where the resistor once was. Label this circuit with $+$ and $-$ polarity markings and the symbol v_{oc}. Then use any circuit analysis technique to determine the value of v_{oc}. In the examples we will usually use the mesh current method or the node voltage method.

2. Next calculate the Thévenin equivalent resistance if the circuit contains only independent sources and resistors, or the short-circuit current if the circuit also contains dependent sources. To calculate the Thévenin equivalent resistance, draw the circuit with the load resistor removed. From the perspective of the resulting open circuit, calculate the equivalent resistance. To do this, replace all voltage sources with short circuits and all current sources with open circuits. Then make series and parallel combinations of the remaining resistors until only one resistor remains. This is the Thévenin equivalent resistor. If there are dependent sources in the circuit, you cannot use the previous method to calculate the Thévenin equivalent resistance because you cannot remove the independent sources without changing the way the dependent sources behave. Therefore you must calculate the short-circuit current instead. To do this, draw the circuit with the load resistor removed and replaced by a short circuit (a wire). Label the current in the short circuit i_{sc}. Then use any circuit analysis technique to determine the value of i_{sc}. We usually employ the node voltage method or the mesh current method. Remember that we can use the open-circuit voltage and the short-circuit current to determine the Thévenin equivalent resistance with the equation

$$R_{Th} = \frac{v_{oc}}{i_{sc}}.$$

3. Now draw the Thévenin equivalent circuit, which consists of a voltage source with the value v_{oc} in series with a resistor whose value is R_{Th}, or the Norton equivalent circuit, which consists of a current source with the value i_{sc} in parallel with a resistor whose value is R_{Th}. Then attach the original load resistor to complete the circuit. Use any circuit analysis method to determine the requested voltage, current or power in this simplified circuit.

4. You can check your result by analyzing the original circuit using any appropriate circuit analysis technique. We will usually employ the node voltage method or the mesh current method.

The first example is a circuit without dependent sources. We consider circuits with dependent sources in the second example.

Example 5.1

Find v_o for the circuit in Fig. 5.1 by replacing the circuit to the left of the 15Ω resistor with its Thévenin equivalent and analyzing the resulting simplified circuit.

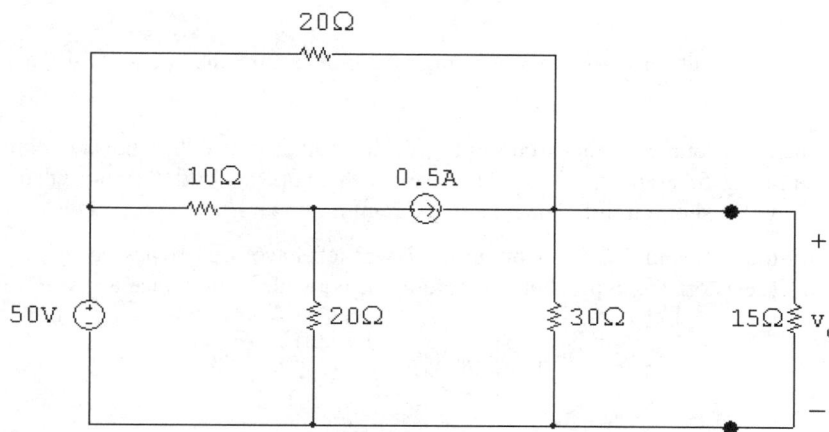

Figure 5.1: The circuit for Example 5.1

Solution

1. Redraw the circuit in Fig. 5.1 with the 15Ω resistor replaced by an open circuit labeled v_{oc} and calculate the value of v_{oc}. We will use the node voltage method to determine v_{oc}, so we have identified the reference node and labeled the remaining non-reference essential nodes with symbols if the voltage at the node is not known. The resulting circuit is shown in Fig. 5.2.

The node voltage equations are

$$\text{At } v_1: \quad \frac{v_1 - 50}{10} + \frac{v_1}{20} + 0.5 = 0$$

$$\text{At } v_{oc}: \quad \frac{v_{oc} - 50}{20} + \frac{v_{oc}}{30} - 0.5 = 0$$

Rewriting the node voltage equations in standard form we get

$$\text{At } v_1: \quad \left(\frac{1}{10} + \frac{1}{20}\right) v_1 + (0)v_{oc} = (50/10) - 0.5$$

$$\text{At } v_{oc}: \quad (0)v_1 + \left(\frac{1}{20} + \frac{1}{30}\right) v_{oc} = (50/20) + 0.5$$

The calculator solution is

$$v_1 = 30 \text{ V}; \qquad v_{oc} = 36 \text{ V}.$$

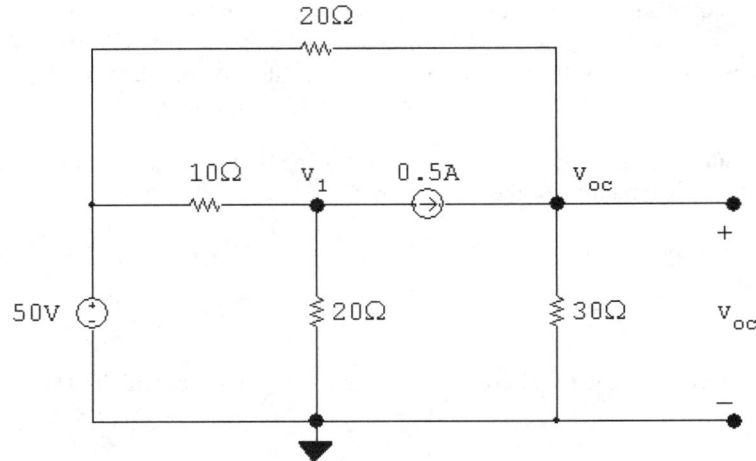

Figure 5.2: The circuit for Example 5.1, configured to determine the open-circuit voltage v_{oc}.

2. Since there are no dependent sources in the circuit in Fig. 5.1 we can calculate the Thévenin equivalent resistance. To do this, redraw the circuit in Fig. 5.1, replacing the 15Ω resistor with an open circuit, the current source with an open circuit, and the voltage source with a short circuit. The resulting circuit is shown in Fig. 5.3.

Notice that in Fig. 5.3 the 10Ω and 20Ω resistors in the lower left have been bypassed by a short circuit, and that the remaining 20Ω and 30Ω resistors are in parallel. Therefore, the equivalent resistance is given by

$$R_{\text{Th}} = 20\|30 = \frac{(20)(30)}{20 + 30} = 12\Omega.$$

3. Now draw the Thévenin equivalent circuit and attach the 15Ω resistor. The result is shown in Fig. 5.4.

To find v_o in this simple circuit, use voltage division:

$$v_o = \frac{15}{15 + 12}(36) = 20 \text{ V}$$

4. We can check this result by analyzing the original circuit in Fig. 5.1 to find v_o. We choose the node voltage method for this analysis, and the circuit in Fig. 5.5 is configured for such analysis.

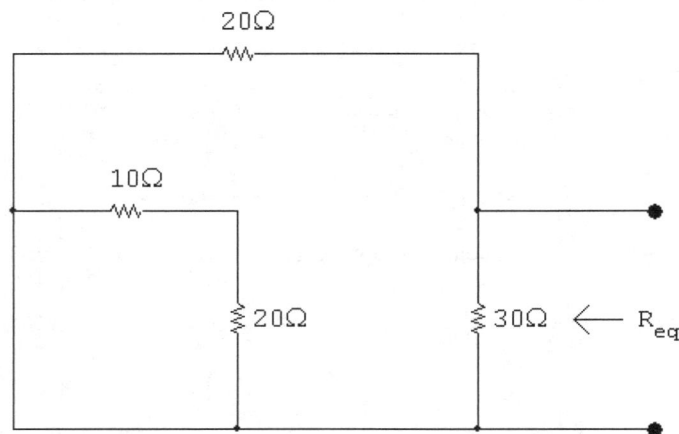

Figure 5.3: The circuit for Example 5.1, configured to determine the Thévenin equivalent resistance R_{Th}.

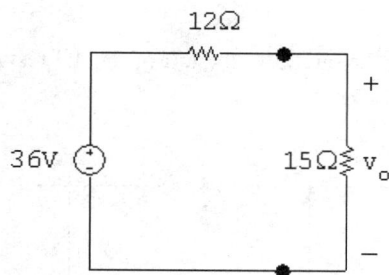

Figure 5.4: The circuit for Example 5.1, with the components to the left of the 15Ω resistor replaced by a Thévenin equivalent.

The node voltage equations are

$$\text{At } v_1: \quad \frac{v_1 - 50}{10} + \frac{v_1}{20} + 0.5 = 0$$

$$\text{At } v_o: \quad \frac{v_o - 50}{20} + \frac{v_o}{30} + \frac{v_o}{15} - 0.5 = 0$$

Rewriting the node voltage equations in standard form we get

$$\text{At } v_1: \quad \left(\frac{1}{10} + \frac{1}{20}\right)v_1 + (0)v_{oc} = (50/10) - 0.5$$

$$\text{At } v_o: \quad (0)v_1 + \left(\frac{1}{20} + \frac{1}{30} + \frac{1}{15}\right) = (50/20) + 0.5$$

The calculator solution is

$$v_1 = 30 \text{ V}; \qquad v_o = 20 \text{ V}.$$

Thus the solution obtained with the Thévenin equivalent circuit is confirmed.

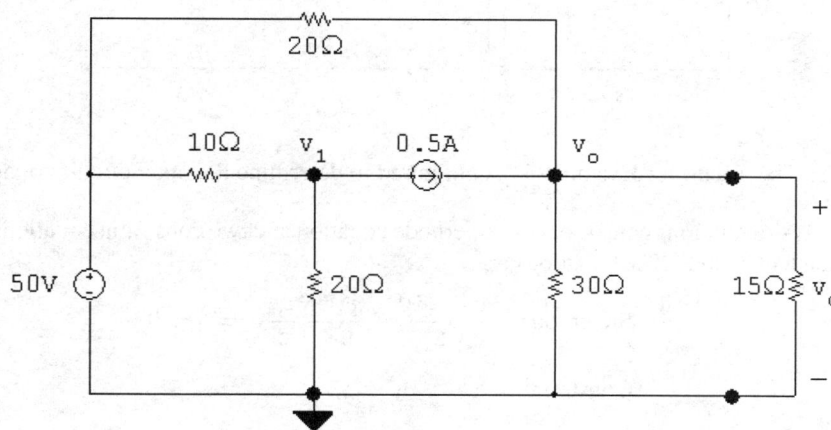

Figure 5.5: The circuit for Example 5.1, prepared for node voltage analysis.

Example 5.2

Find i_o for the circuit in Fig. 5.6 by replacing the circuit to the left of the 4Ω resistor with its Norton equivalent and analyzing the resulting simplified circuit.

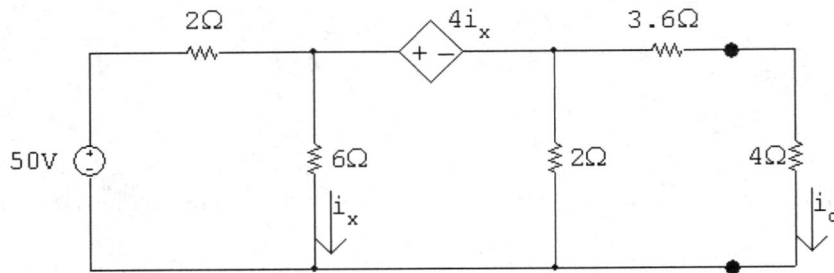

Figure 5.6: The circuit for Example 5.2

Solution

1. Redraw the circuit in Fig. 5.6 with the 4Ω resistor replaced by an open circuit labeled v_{oc} and calculate the value of v_{oc}. We will use the node voltage method to determine v_{oc}, so we have identified the reference node. The remaining non-reference essential nodes form a single supernode with the dependent source, so we label those nodes with symbols and identify the suprenode. The resulting circuit is shown in Fig. 5.7.

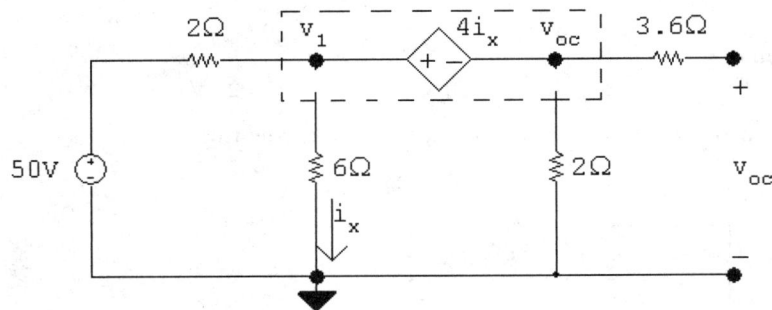

Figure 5.7: The circuit for Example 5.2, configured to determine the open-circuit voltage v_{oc}.

The node voltage analysis equations consist of one supernode equation and two constraint equations, one for the supernode and one for the dependent source. The equations are

$$\text{Supernode:} \quad \frac{v_1 - 50}{2} + \frac{v_1}{6} + \frac{v_{oc}}{2} = 0$$

$$\text{Constraint:} \quad v_1 - v_{oc} = 4i_x$$

$$\text{Constraint:} \quad \frac{v_1}{6} = i_x$$

Rewriting the node voltage equations in standard form we get

$$\text{Supernode:} \quad \left(\frac{1}{2}+\frac{1}{6}\right)v_1 + \left(\frac{1}{2}\right)v_{oc} + (0)i_x = (50/2)$$
$$\text{Constraint:} \quad (1)v_1 + (-1)v_{oc} + (-4)i_x = 0$$
$$\text{Constraint:} \quad \left(\frac{1}{6}\right)v_1 + (0)v_{oc} + (-1)i_x = 0$$

The calculator solution is
$$v_1 = 30 \text{ V}; \qquad v_{oc} = 10 \text{ V}; \qquad i_x = 5 \text{ A}.$$

2. Since there is a dependent source in the circuit in Fig. 5.6 we must calculate the short-circuit current. To do this, redraw the circuit in Fig. 5.6, replacing the 4Ω resistor with short circuit and label the current in the short circuit i_{sc}. Since we want to find this short-circuit current the mesh current method is a good choice, so we also identify and label the mesh currents. The resulting circuit is shown in Fig. 5.8.

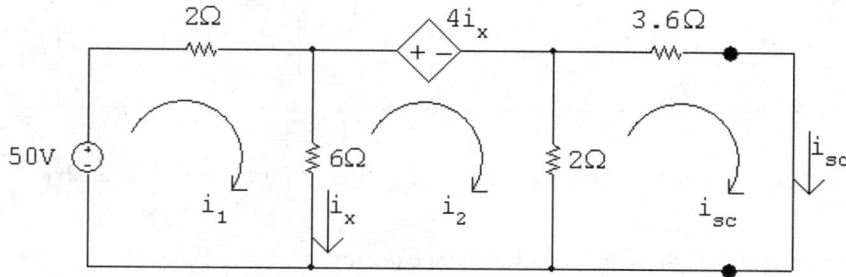

Figure 5.8: The circuit for Example 5.2, configured to determine the short-circuit current i_{sc}.

We need three mesh current equations and a constraint equation for the dependent source. The equations are The equations are

$$
\begin{aligned}
\text{Left mesh:} \quad & -50 + 8i_1 + 6(i_1 - i_2) &= 0 \\
\text{Center mesh:} \quad & 4i_x + 2(i_2 - i_{sc}) + 6(i_2 - i_1) &= 0 \\
\text{Right mesh:} \quad & 3.6i_{sc} + 2(i_{sc} - i_2) &= 0 \\
\text{Constraint:} \quad & i_1 - i_2 &= i_x
\end{aligned}
$$

Rewriting the mesh current equations in standard form we get

$$
\begin{aligned}
\text{Left mesh:} \quad & (8)i_1 &+& (-6)i_2 &+& (0)i_{sc} &+& (0)i_x &=& 50 \\
\text{Center mesh:} \quad & (-6)i_1 &+& (8)i_2 &+& (-2)i_{sc} &+& (4)i_x &=& 0 \\
\text{Right mesh:} \quad & (0)i_1 &+& (-2)i_2 &+& (5.6)i_{sc} &+& (0)i_x &=& 0 \\
\text{Constraint:} \quad & (1)i_1 &+& (-1)i_2 &+& (0)i_{sc} &+& (-1)i_x &=& 0
\end{aligned}
$$

The calculator solution is

$$
i_1 = 11.5 \text{ A}; \qquad i_2 = 7 \text{ A}; \qquad i_{sc} = 2.5 \text{ A}; \qquad i_x = 4.5 \text{ A}.
$$

3. Now draw the Norton equivalent circuit by placing a current source whose value is $i_{sc} = 2.5$A in parallel with a resistor whose value is $r_{Th} = v_{oc}/i_{sc} = 10/2.5 = 4\Omega$, attach the 4Ω load resistor. The result is shown in Fig. 5.9.

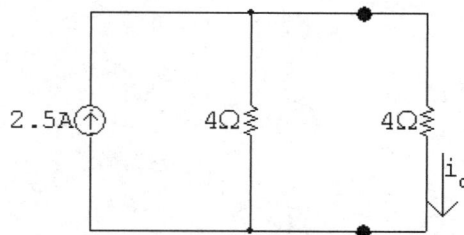

Figure 5.9: The circuit for Example 5.2, with the components to the left of the 4Ω resistor replaced by a Norton equivalent.

To find i_o in this simple circuit, use current division:

$$
i_o = \frac{4}{4+4}(2.5) = 1.25 \text{ A}
$$

4. We can check this result by analyzing the original circuit in Fig. 5.6 to find i_o. We choose the mesh current method for this analysis, and the circuit in Fig. 5.10 is configured for such analysis.

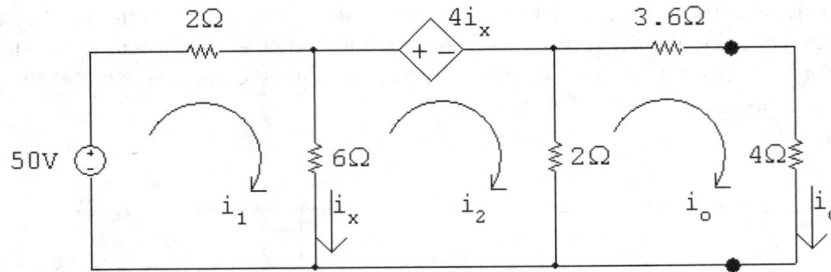

Figure 5.10: The circuit for Example 5.2, prepared for mesh current analysis.

We need three mesh current equations and one constraint equation, as shown below:

$$
\begin{aligned}
\text{Left mesh:} && -50 + 8i_1 + 6(i_1 - i_2) &= 0 \\
\text{Center mesh:} && 4i_x + 2(i_2 - i_o) + 6(i_2 - i_1) &= 0 \\
\text{Right mesh:} && 3.6i_o + 4i_o + 2(i_o - i_2) &= 0 \\
\text{Constraint:} && i_1 - i_2 &= i_x
\end{aligned}
$$

Rewriting the mesh current equations in standard form we get

$$
\begin{aligned}
\text{Left mesh:}   && (8)i_1   &+ (-6)i_2 &+ (0)i_o   &+ (0)i_x   &= 50 \\
\text{Center mesh:} && (-6)i_1 &+ (8)i_2 &+ (-2)i_o &+ (4)i_x   &= 0 \\
\text{Right mesh:} && (0)i_1   &+ (-2)i_2 &+ (9.6)i_o &+ (0)i_x   &= 0 \\
\text{Constraint:} && (1)i_1   &+ (-1)i_2 &+ (0)i_o   &+ (-1)i_x &= 0
\end{aligned}
$$

The calculator solution is

$$i_1 = 10.75 \text{ A}; \qquad i_2 = 6 \text{ A}; \qquad i_o = 1.25 \text{ A}; \qquad i_x = 4.75 \text{ A}.$$

Thus the solution obtained with the Norton equivalent circuit is confirmed.

Now try using the Thévenin equivalent method for each of the practice problems below.

Practice Problem 5.1

Find v_o for the circuit in Fig. 5.11 by replacing the circuit to the left of the 36Ω resistor with its Thévenin equivalent.

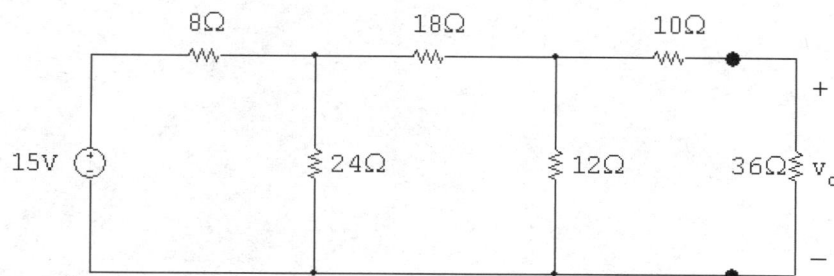

Figure 5.11: The circuit for Practice Problem 5.1.

1. Redraw the circuit in Fig. 5.11, replacing the 36Ω resistor with an open circuit. Use this circuit to calculate v_{oc}.

2. Are there dependent sources in the circuit? If not, find the Thévenin equivalent resistor by redrawing the circuit in Fig. 5.11 with the load resistor removed, the voltage sources replaced by short circuits, and the current sources replaced with open circuits. Then make series and parallel combinations of resistors until a single equivalent resistor remains. If there are dependent sources in the circuit, find the short circuit current by redrawing the circuit in Fig. 5.11, replacing the 36Ω resistor with a short circuit whose current is i_{sc}. Use this circuit to find i_{sc}.

3. Draw the Thévenin equivalent circuit and attach the 36Ω resistor. Use this circuit to calculate v_o.

4. Check your solution by analyzing the original circuit in Fig. 5.11 to find v_o.

Practice Problem 5.2

Find i_o for the circuit in Fig. 5.12 by replacing the circuit to the left of the 3Ω resistor with its Norton equivalent.

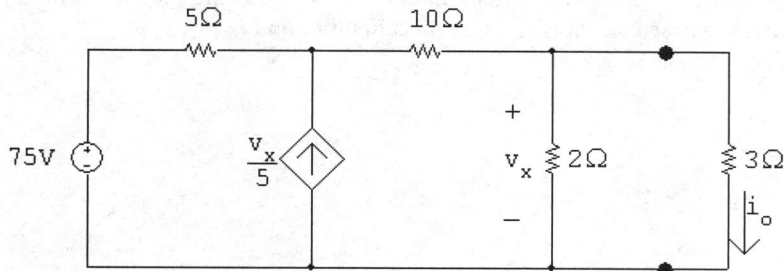

Figure 5.12: The circuit for Practice Problem 5.2.

1. Redraw the circuit in Fig. 5.12, replacing the 3Ω resistor with an open circuit. Use this circuit to calculate v_{oc}.

2. Are there dependent sources in the circuit? If not, find the Thévenin equivalent resistor by redrawing the circuit in Fig. 5.12 with the load resistor removed, the voltage sources replaced by short circuits, and the current sources replaced with open circuits. Then make series and parallel combinations of resistors until a single equivalent resistor remains. If there are dependent sources in the circuit, find the short circuit current by redrawing the circuit in Fig. 5.12, replacing the load resistor with a short circuit whose current is i_{sc}. Use this circuit to find i_{sc}.

3. Draw the Norton equivalent circuit and attach the 3Ω resistor. Use this circuit to calculate i_o.

4. Check your solution by analyzing the original circuit in Fig. 5.12 to find i_o.

Practice Problem 5.3

Find v_o for the circuit in Fig. 5.13 by replacing the circuit to the left of the 16Ω resistor with its Thévenin equivalent.

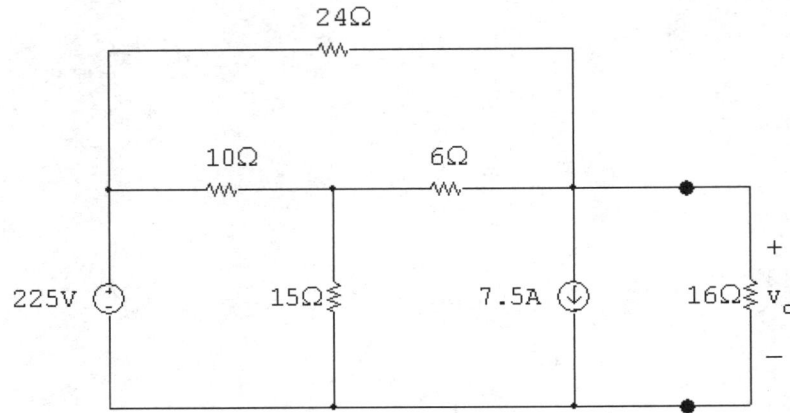

Figure 5.13: The circuit for Practice Problem 5.3.

1. Redraw the circuit in Fig. 5.13, replacing the 16Ω resistor with an open circuit. Use this circuit to calculate v_{oc}.

2. Are there dependent sources in the circuit? If not, find the Thévenin equivalent resistor by redrawing the circuit in Fig. 5.13 with the load resistor removed, the voltage sources replaced by short circuits, and the current sources replaced with open circuits. Then make series and parallel combinations of resistors until a single equivalent resistor remains. If there are dependent sources in the circuit, find the short circuit current by redrawing the circuit in Fig. 5.13, replacing the load resistor with a short circuit whose current is i_{sc}. Use this circuit to find i_{sc}.

3. Draw the Thévenin equivalent circuit and attach the 16Ω resistor. Use this circuit to calculate v_o.

4. Check your solution by analyzing the original circuit in Fig. 5.13 to find v_o.

Practice Problem 5.4

Find power dissipated in the 40Ω resistor for the circuit in Fig. 5.14 by replacing the circuit to the left of the 40Ω resistor with its Thévenin equivalent.

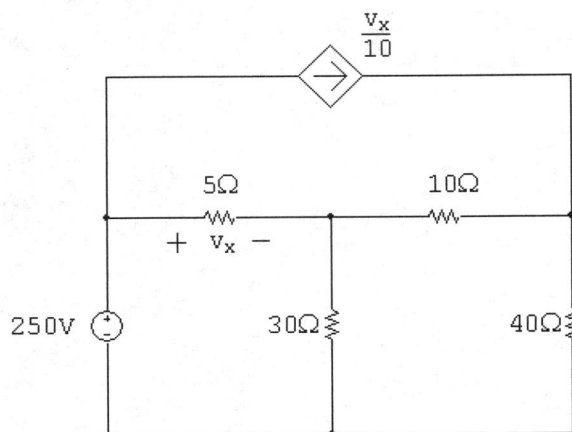

Figure 5.14: The circuit for Practice Problem 5.4.

1. Redraw the circuit in Fig. 5.14, replacing the 40Ω resistor with an open circuit. Use this circuit to calculate v_{oc}.

2. Are there dependent sources in the circuit? If not, find the Thévenin equivalent resistor by redrawing the circuit in Fig. 5.14 with the load resistor removed, the voltage sources replaced by short circuits, and the current sources replaced with open circuits. Then make series and parallel combinations of resistors until a single equivalent resistor remains. If there are dependent sources in the circuit, find the short circuit current by redrawing the circuit in Fig. 5.14, replacing the load resistor with a short circuit whose current is i_{sc}. Use this circuit to find i_{sc}.

3. Draw the Thévenin equivalent circuit and attach the 40Ω resistor. Use this circuit to calculate the power dissipated by this 40Ω resistor.

4. Check your solution by analyzing the original circuit in Fig. 5.14 to find $p_{40\Omega}$.

Practice Problem 5.5

Find v_o for the circuit in Fig. 5.15 by replacing the circuit to the left of the 16Ω resistor with its Thévenin equivalent.

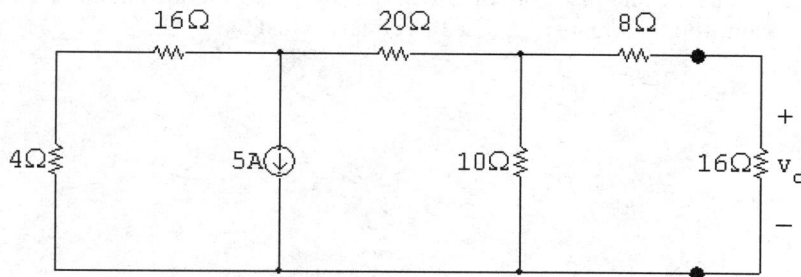

Figure 5.15: The circuit for Practice Problem 5.5.

1. Redraw the circuit in Fig. 5.15, replacing the 16Ω resistor with an open circuit. Use this circuit to calculate v_{oc}.

2. Are there dependent sources in the circuit? If not, find the Thévenin equivalent resistor by redrawing the circuit in Fig. 5.15 with the load resistor removed, the voltage sources replaced by short circuits, and the current sources replaced with open circuits. Then make series and parallel combinations of resistors until a single equivalent resistor remains. If there are dependent sources in the circuit, find the short circuit current by redrawing the circuit in Fig. 5.15, replacing the load resistor with a short circuit whose current is i_{sc}. Use this circuit to find i_{sc}.

3. Draw the Thévenin equivalent circuit and attach the 16Ω resistor. Use this circuit to calculate v_o.

4. Check your solution by analyzing the original circuit in Fig. 5.15 to find v_o.

Practice Problem 5.6

Find the power dissipated in the 100Ω resistor for the circuit in Fig. 5.16 by replacing the circuit to the left of the 100Ω resistor with its Thévenin equivalent.

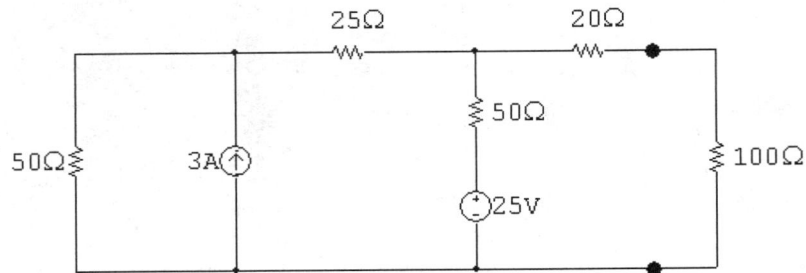

Figure 5.16: The circuit for Practice Problem 5.6.

1. Redraw the circuit in Fig. 5.16, replacing the 100Ω resistor with an open circuit. Use this circuit to calculate v_{oc}.

2. Are there dependent sources in the circuit? If not, find the Thévenin equivalent resistor by redrawing the circuit in Fig. 5.16 with the load resistor removed, the voltage sources replaced by short circuits, and the current sources replaced with open circuits. Then make series and parallel combinations of resistors until a single equivalent resistor remains. If there are dependent sources in the circuit, find the short circuit current by redrawing the circuit in Fig. 5.16, replacing the load resistor with a short circuit whose current is i_{sc}. Use this circuit to find i_{sc}.

3. Draw the Thévenin equivalent circuit and attach the 100Ω resistor. Use this circuit to calculate the power dissipated in this resistor.

4. Check your solution by analyzing the original circuit in Fig. 5.16 to find $p_{100\Omega}$.

Practice Problem 5.7

Find v_o for the circuit in Fig. 5.17 by replacing the circuit to the left of the 250Ω resistor with its Thévenin equivalent.

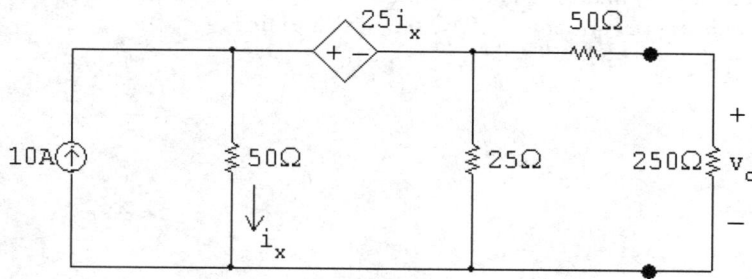

Figure 5.17: The circuit for Practice Problem 5.7.

1. Redraw the circuit in Fig. 5.17, replacing the 250Ω resistor with an open circuit. Use this circuit to calculate v_{oc}.

2. Are there dependent sources in the circuit? If not, find the Thévenin equivalent resistor by redrawing the circuit in Fig. 5.17 with the load resistor removed, the voltage sources replaced by short circuits, and the current sources replaced with open circuits. Then make series and parallel combinations of resistors until a single equivalent resistor remains. If there are dependent sources in the circuit, find the short circuit current by redrawing the circuit in Fig. 5.17, replacing the load resistor with a short circuit whose current is i_{sc}. Use this circuit to find i_{sc}.

3. Draw the Thévenin equivalent circuit and attach the 250Ω resistor. Use this circuit to calculate v_o.

4. Check your solution by analyzing the original circuit in Fig. 5.17 to find v_o.

Practice Problem 5.8

Find i_o for the circuit in Fig. 5.18 by replacing the circuit to the left of the 80Ω resistor with its Norton equivalent.

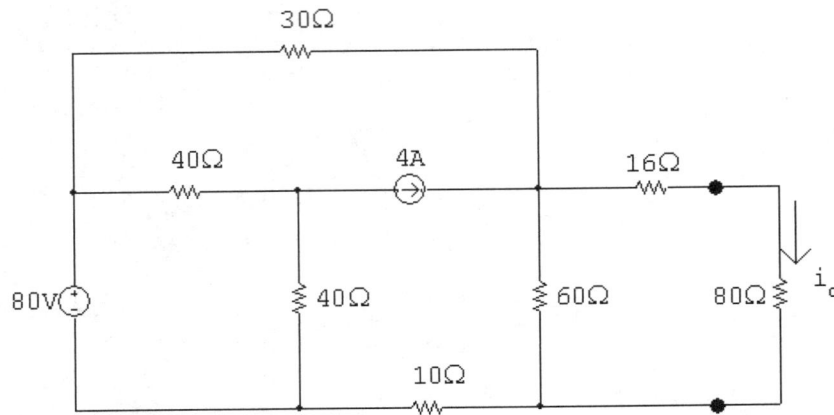

Figure 5.18: The circuit for Practice Problem 5.8.

1. Redraw the circuit in Fig. 5.18, replacing the 80Ω resistor with an open circuit. Use this circuit to calculate v_{oc}.

2. Are there dependent sources in the circuit? If not, find the Thévenin equivalent resistor by redrawing the circuit in Fig. 5.18 with the load resistor removed, the voltage sources replaced by short circuits, and the current sources replaced with open circuits. Then make series and parallel combinations of resistors until a single equivalent resistor remains. If there are dependent sources in the circuit, find the short circuit current by redrawing the circuit in Fig. 5.18, replacing the load resistor with a short circuit whose current is i_{sc}. Use this circuit to find i_{sc}.

3. Draw the Norton equivalent circuit and attach the 80Ω resistor. Use this circuit to calculate i_o.

4. Check your solution by analyzing the original circuit in Fig. 5.18 to find i_o.

Reading

- in *Electric Circuits*, ninth edition:
    - ◆ Section 4.10 — Thévenin and Norton equivalents
    - ◆ Section 4.11 — more Thévenin and Norton equivalents
- Workbook section — Node Voltage Method
- Workbook section — Mesh Current Method

Additional Problems

- 4.63 – 4.68
- 4.73 – 4.74
- 4.77 – 4.78

Solutions

- Practice Problem 5.1:

$$v_{\text{oc}} = 3.75 \text{ V} \qquad i_{\text{sc}} = 208.33 \text{ mA} \qquad R_{\text{Th}} = 18\Omega \qquad v_o = 2.5 \text{ V}.$$

- Practice Problem 5.2:

$$v_{\text{oc}} = 10 \text{ V} \qquad i_{\text{sc}} = 5 \text{ A} \qquad R_{\text{Th}} = 2\Omega \qquad i_o = 2 \text{ A}.$$

- Practice Problem 5.3:

$$v_{\text{oc}} = 105 \text{ V} \qquad i_{\text{sc}} = 13.125 \text{ A} \qquad R_{\text{Th}} = 8\Omega \qquad v_o = 70 \text{ V}.$$

- Practice Problem 5.4:

$$v_{\text{oc}} = 250 \text{ V} \qquad i_{\text{sc}} = 25 \text{ A} \qquad R_{\text{Th}} = 10\Omega \qquad p_{40\Omega} = 1 \text{ kW}.$$

- Practice Problem 5.5:

$$v_{\text{oc}} = -20 \text{ V} \qquad i_{\text{sc}} = -1.25 \text{ A} \qquad R_{\text{Th}} = 16\Omega \qquad v_o = -10 \text{ V}.$$

- Practice Problem 5.6:

$$v_{\text{oc}} = 75 \text{ V} \qquad i_{\text{sc}} = 1.5 \text{ A} \qquad R_{\text{Th}} = 50\Omega \qquad p_{100\Omega} = 25 \text{ W}.$$

- Practice Problem 5.7:

$$v_{\text{oc}} = 125 \text{ V} \qquad i_{\text{sc}} = 2 \text{ A} \qquad R_{\text{Th}} = 62.5\Omega \qquad v_o = 100 \text{ V}.$$

- Practice Problem 5.8:

$$v_{\text{oc}} = 120 \text{ V} \qquad i_{\text{sc}} = 3 \text{ A} \qquad R_{\text{Th}} = 40\Omega \qquad i_o = 1 \text{ A}.$$

Chapter 6

Circuits with Op Amps

The **operational amplifier (op amp)** is a complex nonlinear device with three distinct operating regions: a linear region, in which the output voltage is proportional to the difference between the two input voltages, and two saturation regions where the output voltage takes on either the positive power supply voltage or the negative power supply voltage. But with two simplifying assumptions about the behavior and characteristics of op amps we can readily analyze circuits containing these devices. The assumptions are as follows:

- The op amp is ideal. From the standpoint of the op amp as a device, this means that the op amp has infinite input resistance, infinite open loop gain, and zero output resistance. From the standpoint of analyzing a circuit containing an op amp, this means that there is no current flowing into the input terminals of the op amp (because of the infinite input resistance) and there is no voltage drop across the input terminals of the op amp when it is operating in its linear region (because of the infinite open loop gain).

- The op amp is operating in its linear region. In order to make this assumption, the circuit containing the op amp must have a **negative feedback path** which is a connection from the output terminal of the op amp to the inverting input terminal (the one labeled with a − sign). This assumption, combined with the assumption that the op amp is ideal, allows us to assume that there is no difference between the voltages at the input terminals of the op amp. This assumption leads to the conclusion that the output voltage must be within the range of voltage values established by the positive and negative power supplies if the op amp is indeed within its linear operating region.

To analyze a circuit with an op amp, we begin by making the above assumptions. Then we write one or more node voltage equations at the input terminals of the op amp. Note that we can never write a node voltage equation at the output terminal of the op amp, because we have no method for calculating the current flowing into the op amp at the output terminal. Once we solve the node voltage equations, we check to see whether or not our second assumption can be validated. To do this, we check the voltage at the output of the op amp to see whether or not its value is within the range of values established by the positive and negative power supplies. If the output voltage is within the specified range, our analysis is complete; if not, the output voltage saturates at the power supply voltage closest to the one calculated in our analysis.

We divide the analysis of a circuit containing an op amp into four steps:

1. Assume that the op amp is ideal and operating in its linear region. Label the two input nodes for the op amp with voltages, usually v_p for the non-inverting terminal (the one with the + sign) and v_n for the inverting terminal (the one with the − sign). Label the output node for the op amp with a voltage, usually v_o.

2. If possible, calculate the numerical value of the node voltage at the non-inverting input to the op amp. Remember that the ideal op amp assumption tells us that there is no current flowing into the op amp. If a numerical calculation is not possible, calculate the node voltage at the non-inverting terminal as a function of the source voltage or voltages connected to that terminal

3. Now, write a KCL equation at the inverting input terminal of the op amp. Remember that by assumption, the voltage at the inverting input node is the same as the voltage at the non-inverting input node calculated in Step 2. The node voltage equation written at the inverting terminal will always involve the output voltage variable because of the negative feedback path that allows the op amp to operate within its linear region. Then solve the node voltage equation for the voltage at the output node.

4. Examine the value of the voltage at the output node. If the op amp is actually operating within its linear region, the output voltage will be between the two power supply voltages. If it is, your analysis is complete. If it is not, then the output voltage is not the value you calculated, but instead will saturate at the power supply voltage it is closest to, giving the correct value for the output voltage.

We illustrate this method with the two examples that follow.

Example 6.1

Find the voltage drop v_o for the circuit in Fig. 6.1.

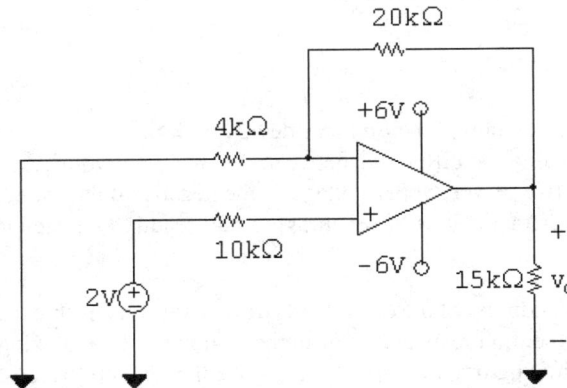

Figure 6.1: The circuit for Example 6.1

Solution

1. We assume that the op amp is ideal and operating in its linear region. This allows us to assume that the value of the current flowing into the op amp at its two input terminals is zero, and that the voltage drop between those same two input terminals is zero. We label the three node voltages, two at the op amp's input and one at its output, as shown in Fig. 6.2.

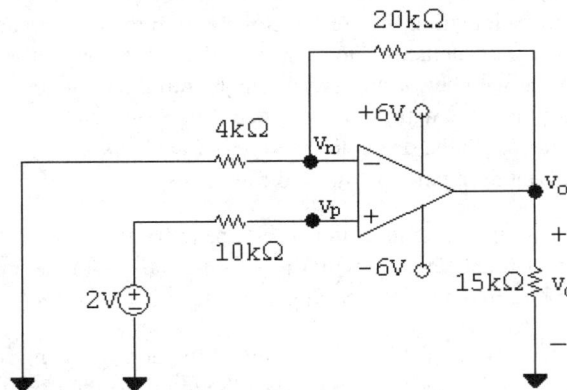

Figure 6.2: The circuit for Example 6.1, with the three node voltages for the op amp identified and labeled.

2. Calculate the value of the voltage at the non-inverting input node of the op amp. We can do this for the voltage v_p quite easily. Since there is no current flowing into the op amp by assumption, there can be no voltage drop across the 10kΩ resistor. Thus,

$$v_p = 2 \text{ V.}$$

3. Write a node voltage equation at the inverting input node of the op amp. The node voltage equation is written by summing the currents leaving the node v_n. Remember that the current leaving this node and flowing into the op amp is zero by assumption.

$$\frac{v_n - 0}{4000} + \frac{v_n - v_o}{20,000} = 0$$

By assumption, there is no voltage drop between the two input terminals for the op amp. Thus,

$$v_n = v_p = 2 \text{ V}.$$

Substituting this value into the node voltage equation and solving for v_o we get

$$v_o = 12 \text{ V}.$$

4. Examine the value of the output voltage. If the op amp is within its linear region of operation, as we assumed, then

$$-6 \text{ V} \le v_o \le +6 \text{ V}.$$

But our calculation gave the result $v_o = 12$ V. Therefore, the assumption of linear operation is invalid, and in fact, the op amp has saturated. The value of the output voltage is the same as the value of the power supply closest to the value of 12 V. Thus,

$$v_o = 6 \text{ V}.$$

Example 6.2

Find the range of values for the voltage v_s such that the output voltage v_o does not saturate for the circuit in Fig. 6.3.

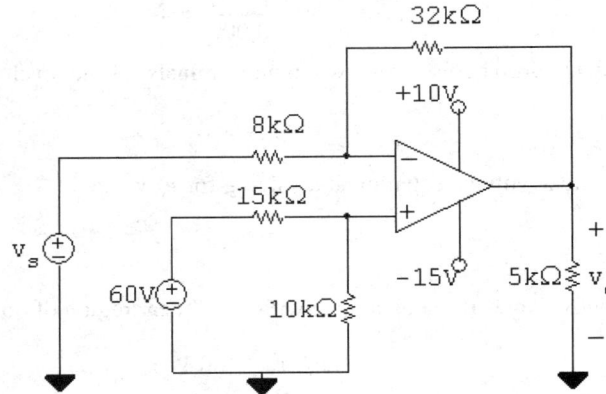

Figure 6.3: The circuit for Example 6.2

Solution

1. We assume that the op amp is ideal and operating in its linear region. This allows us to assume that the value of the current flowing into the op amp at its two input terminals is zero, and that the voltage drop between those same two input terminals is zero. We label the three node voltages, two at the op amp's input and one at its output, as shown in Fig. 6.4.

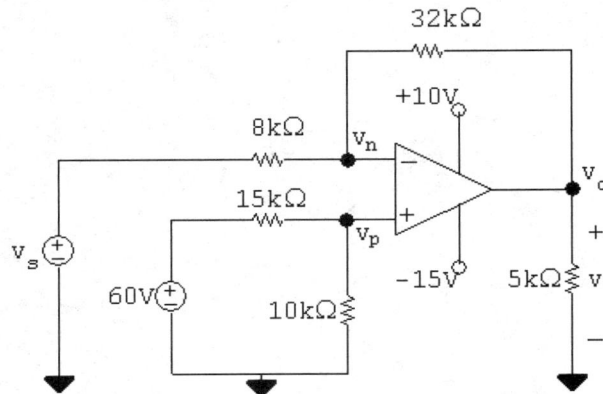

Figure 6.4: The circuit for Example 6.2, with the three node voltages for the op amp identified and labeled.

2. Calculate the value of the voltage at the non-inverting input node of the op amp. We can do this for the voltage v_p using voltage division. Since there is no current flowing into the op amp by assumption, the wire connecting the node labeled v_p to the op amp acts like an open circuit. Therefore, the loop formed by the 60V source, the 15kΩ resistor and the 10kΩ resistor acts as though it is not attached to the rest of the circuit. The voltage v_p is then the voltage drop across the 10kΩ resistor, whose value we calculate using voltage division. Thus,

$$v_p = \frac{10,000}{10,000 + 15,000}(60) = 24 \text{ V.}$$

3. Write a node voltage equation at the inverting input node of the op amp. The node voltage equation is written by summing the currents leaving the node v_n. Remember that the current leaving this node and flowing into the op amp is zero by assumption.

$$\frac{v_n - v_s}{8000} + \frac{v_n - v_o}{32,000} = 0$$

By assumption, there is no voltage drop between the two input terminals for the op amp. Thus,

$$v_n = v_p = 24 \text{ V}.$$

Substituting this value into the node voltage equation and solving for v_o we get

$$v_o = 120 - 4v_s$$

4. Now we use the two power supply voltages as the limiting values for v_0. We consider one limiting value at a time by substituting it into the equation from Step 3 and calculating the value of v_s that would produce this limiting value. When $v_o = 10\text{V}$,

$$10 = 120 - 4v_s \qquad \text{so} \qquad v_s = \frac{120 - 10}{4} = 27.5 \text{ V}.$$

When $v_o = -15\text{V}$,

$$-15 = 120 - 4v_s \qquad \text{so} \qquad v_s = \frac{120 + 15}{4} = 33.75 \text{ V}.$$

Thus, the range of values for v_s for which v_o will not saturate (and the op amp remains in its linear operating region) is

$$27.5 \text{ V} \le v_s \le 33.75 \text{ V}.$$

Now you are ready to practice analyzing circuits with op amps in the problems that follow.

Practice Problem 6.1

Find the range of values for the voltage v_s such that the output voltage v_o does not saturate and the op amp remains in its linear region of operation for the circuit in Fig. 6.5.

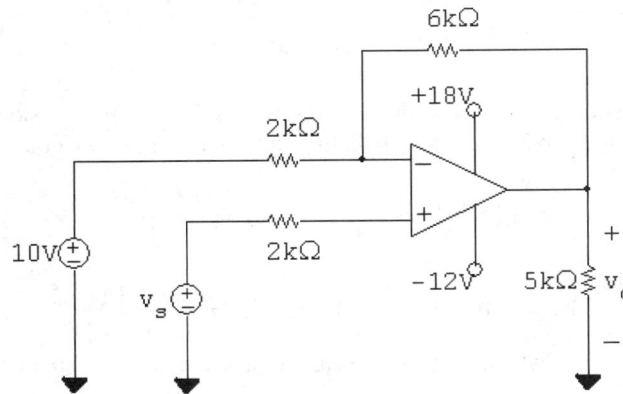

Figure 6.5: The circuit for Practice Problem 6.1.

1. Assume that the op amp is ideal and operating in its linear region. Label the three node voltages, two at the op amp's input and one at its output, in Fig. 6.5.

2. Calculate the value of the voltage at the non-inverting input node of the op amp, or write an equation for the voltage v_p in terms of the source voltage.

3. Write a node voltage equation at the inverting input node of the op amp. Solve this equation for v_o.

4. Use the two power supply voltages as the limiting values for v_o and calculate the range of values for v_s that will keep v_o within its limiting values.

Practice Problem 6.2

Find the range of values for the voltage v_s such that the output voltage v_o does not saturate and the op amp remains in its linear region of operation for the circuit in Fig. 6.6.

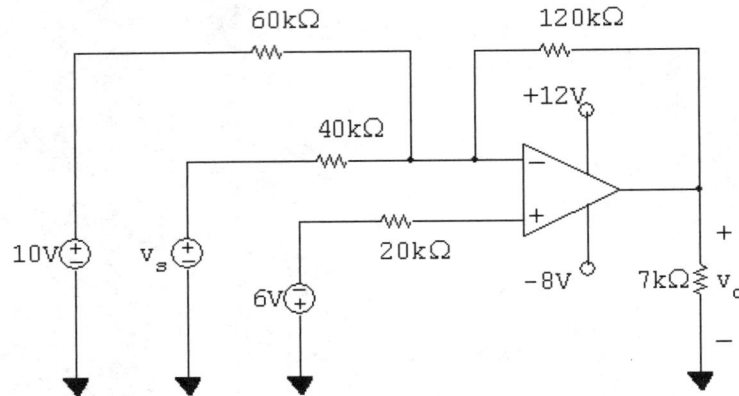

Figure 6.6: The circuit for Practice Problem 6.2.

1. Assume that the op amp is ideal and operating in its linear region. Label the three node voltages, two at the op amp's input and one at its output, in Fig. 6.6.

2. Calculate the value of the voltage at the non-inverting input node of the op amp.

3. Write a node voltage equation at the inverting input node of the op amp. Solve this equation for v_o.

4. Use the two power supply voltages as the limiting values for v_o and calculate the range of values for v_s that will keep v_o within its limiting values.

Practice Problem 6.3

Calculate i_o for the circuit in Fig. 6.7.

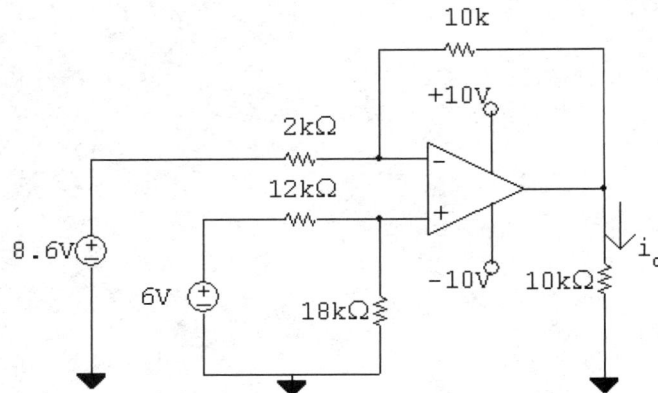

Figure 6.7: The circuit for Practice Problem 6.3.

1. Assume that the op amp is ideal and operating in its linear region. Label the three node voltages, two at the op amp's input and one at its output, in Fig. 6.7.

2. Calculate the value of the voltage at the non-inverting input node of the op amp, or write an equation for the voltage v_p.

3. Write a node voltage equation at the inverting input node of the op amp. Solve this equation for v_o. Use this value of v_o to calculate i_o.

4. Is the value for v_o within the range of voltages defined by the power supplies? If so, your value of i_o is correct. If not, you must recalculate i_o based on the saturated value of v_o.

Practice Problem 6.4

What value for R_f will yield the equation $v_o = 5 - 4v_a$ for the circuit in Fig. 6.8.

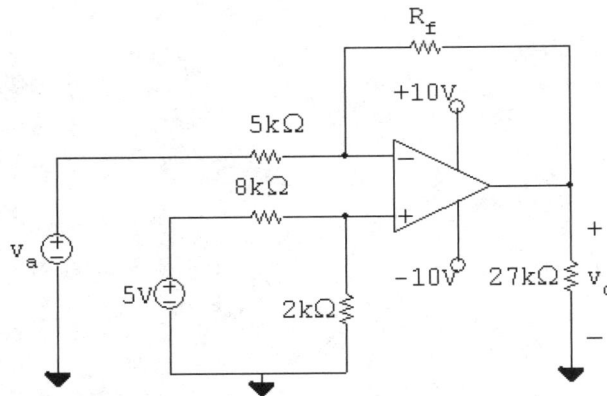

Figure 6.8: The circuit for Practice Problem 6.4.

1. Assume that the op amp is ideal and operating in its linear region. Label the three node voltages, two at the op amp's input and one at its output, in Fig. 6.8.

2. Calculate the value of the voltage at the non-inverting input node of the op amp, or write an equation for the voltage v_p.

3. Write a node voltage equation at the inverting input node of the op amp. Simplify this equation for v_o. Use this equation for v_o to calculate R_f.

4. This problem does not concern a calculated value for v_o. We assume that the op amp is in its linear region of operation in order to obtain the equation specified in the problem.

Practice Problem 6.5

Find the range of values for the voltage v_s such that the output voltage v_o does not saturate and the op amp remains in its linear region of operation for the circuit in Fig. 6.9.

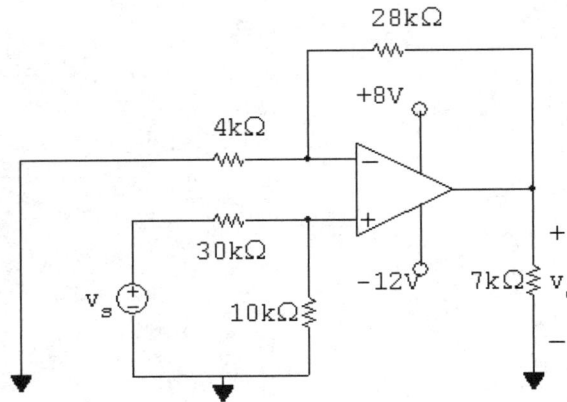

Figure 6.9: The circuit for Practice Problem 6.5.

1. Assume that the op amp is ideal and operating in its linear region. Label the three node voltages, two at the op amp's input and one at its output, in Fig. 6.9.

2. Calculate the value of the voltage at the non-inverting input node of the op amp, or write an equation for the voltage v_p in terms of the source voltage.

3. Write a node voltage equation at the inverting input node of the op amp. Solve this equation for v_o.

4. Use the two power supply voltages as the limiting values for v_o and calculate the range of values for v_s that will keep v_o within its limiting values.

Practice Problem 6.6

Find the range of values for the voltage v_a such that the output voltage v_o does not saturate and the op amp remains in its linear region of operation for the circuit in Fig. 6.10.

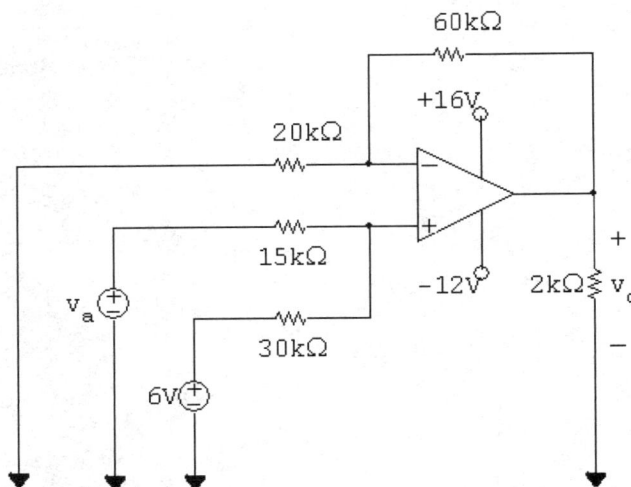

Figure 6.10: The circuit for Practice Problem 6.6.

1. Assume that the op amp is ideal and operating in its linear region. Label the three node voltages, two at the op amp's input and one at its output, in Fig. 6.10.

2. Calculate the value of the voltage at the non-inverting input node of the op amp, or write an equation for the voltage v_p in terms of the source voltage v_a.

3. Write a node voltage equation at the inverting input node of the op amp. Solve this equation for v_o.

4. Use the two power supply voltages as the limiting values for v_o and calculate the range of values for v_a that will keep v_o within its limiting values.

Practice Problem 6.7

Find the range of values for the voltage v_s such that the output voltage v_o does not saturate and the op amp remains in its linear region of operation for the circuit in Fig. 6.11.

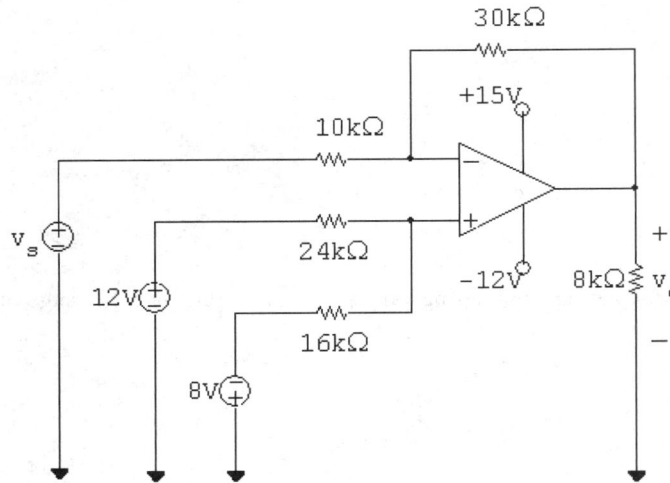

Figure 6.11: The circuit for Practice Problem 6.7.

1. Assume that the op amp is ideal and operating in its linear region. Label the three node voltages, two at the op amp's input and one at its output, in Fig. 6.11.

2. Calculate the value of the voltage at the non-inverting input node of the op amp, or write an equation for the voltage v_p.

3. Write a node voltage equation at the inverting input node of the op amp. Solve this equation for v_o.

4. Use the two power supply voltages as the limiting values for v_o and calculate the range of values for v_s that will keep v_o within its limiting values.

Practice Problem 6.8

Find the range of values for the voltage v_a such that the output voltage v_o does not saturate and the op amp remains in its linear region of operation for the circuit in Fig. 6.12.

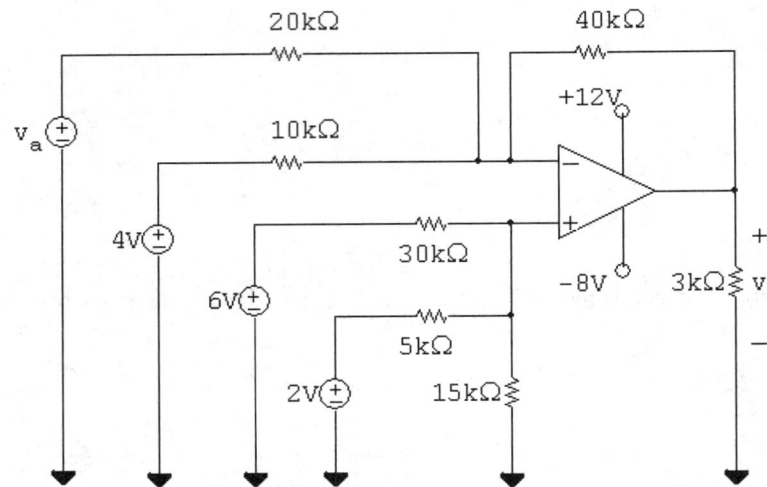

Figure 6.12: The circuit for Practice Problem 6.8.

1. Assume that the op amp is ideal and operating in its linear region. Label the three node voltages, two at the op amp's input and one at its output, in Fig. 6.12.

2. Calculate the value of the voltage at the non-inverting input node of the op amp, or write an equation for the voltage v_p.

3. Write a node voltage equation at the inverting input node of the op amp. Solve this equation for v_o.

4. Use the two power supply voltages as the limiting values for v_o and calculate the range of values for v_a that will keep v_o within its limiting values.

Practice Problem 6.9

Find the range of values for the resistor R_a such that the output voltage v_o does not saturate and the op amp remains in its linear region of operation for the circuit in Fig. 6.13.

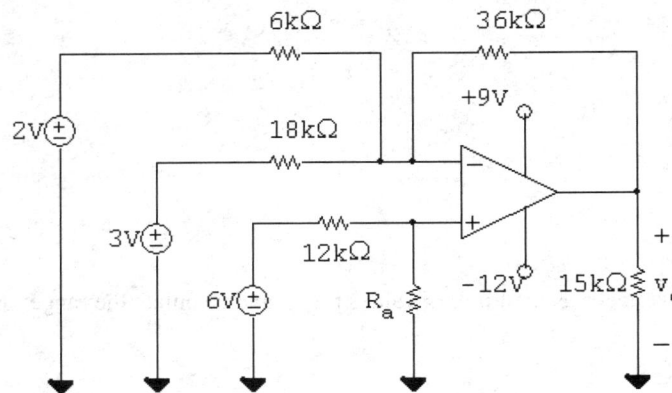

Figure 6.13: The circuit for Practice Problem 6.9.

1. Assume that the op amp is ideal and operating in its linear region. Label the three node voltages, two at the op amp's input and one at its output, in Fig. 6.13.

2. Calculate the value of the voltage at the non-inverting input node of the op amp, or write an equation for the voltage v_p.

3. Write a node voltage equation at the inverting input node of the op amp. Solve this equation for v_o.

4. Use the two power supply voltages as the limiting values for v_o and calculate the range of values for R_f that will keep v_o within its limiting values.

Reading

- in *Electric Circuits*, ninth edition:

  ♦ Section 5.1 — terminology

  ♦ Section 5.2 — op amp operating regions

  ♦ Section 5.3 — inverting amplifier

  ♦ Section 5.4 — summing amplifier

  ♦ Section 5.5 — non-inverting amplifier

  ♦ Section 5.6 — difference amplifier

- Workbook section — Node Voltage Method

Additional Problems

- 5.1

- 5.3

- 5.6

- 5.8 – 5.9

- 5.13 – 5.15

- 5.17 — 5.18

- 5.20 — 5.21

- 5.26

- 5.28

Solutions

- Practice Problem 6.1: $4.5 \text{ V} \leq v_s \leq 12 \text{ V}$.

- Practice Problem 6.2: $-22.67 \text{ V} \leq v_s \leq -16 \text{ V}$.

- Practice Problem 6.3: $i_o = -1 \text{ mA}$.
                            (The op amp output saturates at -10 V.)

- Practice Problem 6.4: $R_f = 20 \text{ k}\Omega$.

- Practice Problem 6.5: $-6 \text{ V} \leq v_s \leq 4 \text{ V}$.

- Practice Problem 6.6: $-7.5 \text{ V} \leq v_a \leq 3 \text{ V}$.

- Practice Problem 6.7: $-5 \text{ V} \leq v_s \leq 4 \text{ V}$.

- Practice Problem 6.8: $-7 \text{ V} \leq v_a \leq 3 \text{ V}$.

- Practice Problem 6.9: $1.5 \text{ k}\Omega \leq R_a \leq 12 \text{ k}\Omega$.

Chapter 7

Natural and Step Response of First-Order (RL and RC) Circuits

Here we review and then practice the techniques that enable us to analyze a particular group of circuits. These are circuits containing one equivalent resistor and either one equivalent inductor or one equivalent capacitor that has initial stored energy. The use of the phrase "one equivalent" means that if the circuit contains two or more resistors, for example, they must be arranged in such a way that they can be combined in series and in parallel to form one single equivalent resistor. The same holds for circuit that contain two or more inductors, or two or more capacitors. These circuits are refered to as RL and RC circuits, and are also called **first-order circuits**, because their describing equation is a first-order differential equation.

These circuits usually contain a switch that is in one position for $t < 0$, switches positions at $t = 0$, and remains at that second position indefinitely. When the switch is in its first position, there is usually an independent current or voltage source in the circuit as well, used to generate the energy that the inductor or capacitor will have stored at $t = 0$. When the switch moves to its second position, there may or may not be an independent current or voltage source in the circuit. If there is, it continues to supply energy to the circuit indefinitely, and we call the analysis a **step response** problem. If there is not an independent source in the circuit for $t \geq 0$, then the energy initially stored is dissipated to the resistor and we call the analysis a **natural response** problem. Fortunately the natural response problem and the step response problem are closely related, so we can use the same circuit analysis technique for both problems.

Analyzing RL circuits is very similar to analyzing RC circuits so we can also use the same circuit analysis technique for both circuits. There are basically three steps in the analysis: find the initial condition, which is the initial current in the inductor and is the initial voltage drop across the capacitor; find the final value, which is the final current in the inductor and is the final voltage drop across the capacitor; and find the **time constant**, τ, which equals L/R for an RL circuit and equals RC for an RC circuit. Once we have these three quantities, the response is given by

$$\text{FV} + (\text{IV} - \text{FV})e^{-t/\tau}$$

where IV is the initial value and FV is the final value. For RL circuits we will use this formula to calculate the current in the inductor; if we want any other quantities, we will calculate them from the inductor current. For RC circuits we will use this formula to calculate the voltage drop across the capacitor; if we want any other quantities, we will again calculate them from the capacitor voltage.

The first-order analysis method for RL and RC circuits can be broken into the following steps:

1. Redraw the circuit as it appears for $t < 0$, replacing the switch with an open circuit if it is open, and with a short circuit if it is closed. Since it is assumed that the switch has been in this position for a long time, this places the inductor or the capacitor in the presence of a constant source. Therefore, an inductor behaves like a short circuit while a capacitor behaves like an open circuit. If you are dealing with an RL circuit, replace the inductor with a short circuit and calculate the current through the short circuit, which will be the initial current, I_o. If you have an RC circuit, replace the capacitor with an open circuit and calculate the voltage drop across the open circuit, which will be the initial voltage, V_o.

2. Redraw the circuit as it appears for $t \geq 0$, replacing the switch with an open circuit if it is open and with a short circuit if it is closed. If there are no independent sources in the circuit, this is the natural response problem and the final value is 0, since all of the initially stored energy in the inductor or capacitor will be dissipated by the resistor as $t \to \infty$. This means that in the RL circuit the final current is $I_f = 0$ and in the RC circuit the final voltage is $V_f = 0$.

If there is an independent source in the circuit for $t \geq 0$ this is the step response problem. If you have an RL circuit, the inductor will have been in the presence of this independent source for a long time as $t \rightarrow \infty$ and so behaves like a short circuit. Replace the inductor with a short circuit and calculate the current in the short circuit. This is the final current, I_f. Make sure the direction of the current arrow is the same when computing I_o in Step 1 and when computing I_f in this step. If you have an RC circuit, the capacitor will have been in the presence of the independent source for a long time as $t \rightarrow \infty$ and behaves like an open circuit. Replace it with an open circuit and calculate the voltage drop across the open circuit. This is the final voltage, V_f. Make sure that the polarity of the voltage when you are calculating the initial voltage V_o is the same as the polarity of the voltage when you are calculating the final value V_f.

3. Redraw the circuit from Step 2 (the circuit as it appears for $t \geq 0$) and make the following modifications to enable you to calculate the time constant, τ. If you have an RL circuit, remove the inductor from the drawing; if you have an RC circuit, remove the capacitor from the circuit. In either case, find the Thévenin equivalent resistance as seen from the open circuit where the component was located. If there are no dependent sources in the circuit, this means you can replace any voltage sources with short circuits, replace any current sources with open circuits, and make series and parallel combinations of the remaining resistors until a single resistor value is found. If there are dependent sources in the circuit you must find the Thévenin equivalent resistance by calculating the open circuit voltage, v_{oc} and the short circuit current, i_{sc} and use the equation $R_{Th} = v_{oc}/i_{sc}$. The Thévenin equivalent resistance is used to find the time constant τ:

$$\text{For } RL \text{ circuits:} \quad \tau = L/R_{Th} \qquad \text{For } RC \text{ circuits:} \quad \tau = R_{Th}C$$

4. Using the initial value from Step 1, the final value from Step 2, and the time constant from Step 3, find the response of the circuit for $t \geq 0$ as follows:

$$\text{For } RL \text{ circuits:} \quad i_L(t) = I_f + (I_o - I_f)e^{-t/\tau}$$

$$\text{For } RC \text{ circuits:} \quad v_C(t) = V_f + (V_o - V_f)e^{-t/\tau}$$

That is, regardless of what currents and voltages in the original circuit are to be calculated, you will always first calculate the current in the inductor for RL circuits and the voltage drop across the capacitor in RC circuits.

5. If a quantity other than the current in the inductor or the voltage drop across the capacitor is requested in the original circuit, use the inductor current or the capacitor voltage, together with other circuit analysis techniques like Ohm's law and Kirchhoff's laws, to calculate the requested quantities.

The following two examples illustrate the process of analyzing first-order circuits.

Example 7.1

Find $i_L(t)$ for the circuit in Fig. 7.1 for $t \geq 0$.

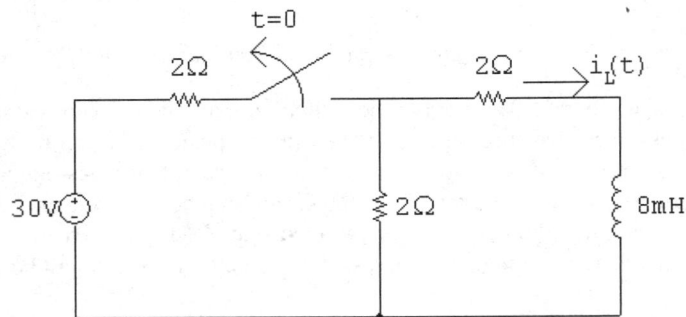

Figure 7.1: The circuit for Example 7.1

Solution

1. Redraw the circuit in Fig. 7.1 with the switch in its closed position. This is the circuit for $t < 0$ and is used to establish the initial condition. It is assumed that the switch has been in this position for a long time, so the inductor has been in the presence of a constant source for a long time and therefore behaves like a short circuit. We thus replace the inductor with a short circuit. Since this is an RL circuit, all of the quantities that we will calculate will be currents. We identify the current in the short circuit as the initial value of the current in the inductor, I_o. The resulting circuit is shown in Fig. 7.2.

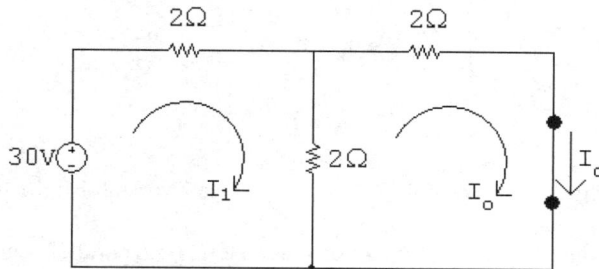

Figure 7.2: The circuit for Example 7.1, for $t < 0$, used to establish the initial condition.

We must analyze this circuit to find I_o. We use the mesh analysis technique, since it will yield I_o directly. The mesh currents are also shown in Fig. 7.2. The mesh current equations are

$$\text{Left mesh:} \quad -30 + 2I_1 + 2(I_1 - I_o) = 0$$
$$\text{Right mesh:} \quad 2I_o + 2(I_o - I_1) = 0$$

Rewriting the mesh current equations in standard form we get

$$\text{Left mesh:} \quad (4)I_1 + (-2)I_o = 30$$
$$\text{Right mesh:} \quad (-2)I_1 + (4)I_o = 0$$

The calculator solution is

$$I_1 = 10 \text{ A}; \qquad I_o = 5 \text{ A}.$$

Thus the initial current $I_o = 5$A.

2. Redraw the circuit in Fig. 7.1 with the switch in its open position. This is the circuit for $t \geq 0$ and is used to establish the final condition, since the circuit will be in this configuration as $t \to \infty$. It is assumed that the switch has been in this position for a long time, so the inductor has been in the presence of a constant source for a long time and therefore behaves like a short circuit. We thus replace the inductor with a short circuit. Since this is an RL circuit, all of the quantities that we will calculate will be currents. We identify the current in the short circuit as the final value of the current in the inductor, I_f. The resulting circuit is shown in Fig. 7.3.

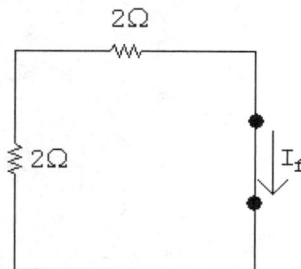

Figure 7.3: The circuit for Example 7.1, for $t \geq 0$, used to establish the final value.

As you can see, there are no independent sources in this circuit, so the stored energy in the inductor will dissipate in the resistors leaving no energy in the inductor. Thus, the final current $I_f = 0$A.

3. To find the time constant for the circuit, we need to find the equivalent resistance seen by the inductor. To do this, redraw the circuit in Fig. 7.3 in Step 2 and replace the short circuit where the inductor was with an open circuit. The resulting circuit is shown in Fig. 7.4.

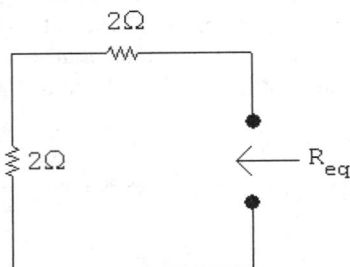

Figure 7.4: The circuit for Example 7.2, for $t \geq 0$, used to establish the time constant, τ.

Since there are no dependent sources in the circuit, we can make series and parallel combinations of the remaining resistors to find the Thévenin equivalent resistance we need for calculating the time constant. As seen in Fig. 7.4, the 2Ω resistors are in series, so

$$R_{Th} = 2 + 2 = 4\Omega.$$

Using this resistance, we can calculate the time constant as follows:

$$\tau = L/R_{Th} = (0.008)/(4) = 2 \text{ ms}.$$

4. Now that we have calculated the initial value of the inductor current, the final value of the inductor current, and the time constant, we can use these values to determine the current in the inductor for $t \geq 0$:

$$i_L(t) = I_f + (I_o - I_f)e^{-t\tau} = 0 + (5 - 0)e^{-t/(0.002)} = 3e^{-500t} \text{ A}, \quad t \geq 0$$

5. Since the current in the inductor was the only quantity requested in the original circuit shown in Fig. 7.1, no further analysis is required.

Example 7.2

Find $i_R(t)$ for the circuit in Fig. 7.5 for $t \geq 0^+$.

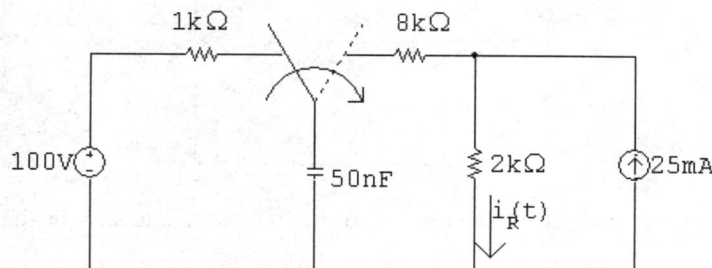

Figure 7.5: The circuit for Example 7.2

Solution

1. Redraw the circuit in Fig. 7.5 with the switch in its left-hand position. This is the circuit for $t < 0$ and is used to establish the initial condition. It is assumed that the switch has been in this position for a long time, so the capacitor has been in the presence of a constant source for a long time and therefore behaves like an open circuit. We thus replace the capacitor with an open circuit. Since this is an RC circuit, all of the quantities that we will calculate will be voltages. We identify the voltage drop across the open circuit as the initial value of the voltage drop across the capacitor, V_o. The resulting circuit is shown in Fig. 7.6.

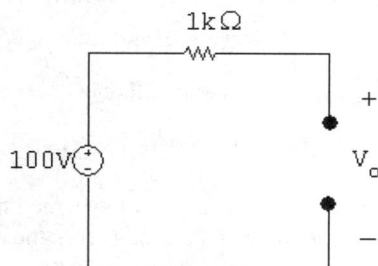

Figure 7.6: The circuit for Example 7.2, for $t < 0$, used to establish the initial condition.

We must analyze this circuit to find V_o. The open circuit that replaced the capacitor will not allow current to flow in the circuit, and thus the voltage drop across the open circuit is fixed by the voltage source. Therefore the initial voltage $V_o = 100$V.

2. Redraw the circuit in Fig. 7.5 with the switch in its right-hand position. This is the circuit for $t \geq 0$ and is used to establish the final condition, since the circuit will be in this configuration as $t \rightarrow \infty$. It is assumed that the switch has been in this position for a long time, so the capacitor has been in the presence of a constant source for a long time and therefore behaves like an open circuit. We thus replace the capacitor with an open circuit. Since this is an RC circuit, all of the quantities that we will calculate will be voltages. We identify the voltage in the open circuit as the final value of the voltage drop across the capacitor, V_f. The resulting circuit is shown in Fig. 7.7.

As you can see, there is an independent source in this circuit, so this is the step response problem. The value of the voltage drop across the capacitor is determined by the independent source and the resistors. Since the capacitor behaves like an open circuit, all of the current from the current source must flow through the 2kΩ resistor, creating a voltage drop of $(2000)(0.025) = 50$V, positive at the top of the open circuit. Thus, the final voltage $V_f = 50$V.

3. To find the time constant for the circuit, we need to find the equivalent resistance seen by the capacitor. To do this, redraw the circuit in Fig. 7.7 in Step 2. Since there are no dependent sources in this circuit we replace the current source with an open circuit. The resulting circuit is shown in Fig. 7.8.

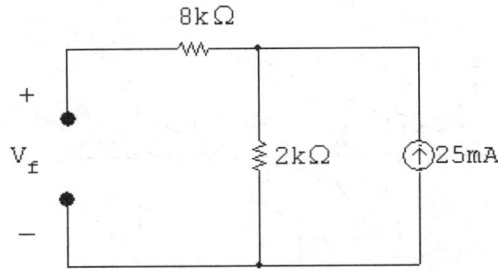

Figure 7.7: The circuit for Example 7.2, for $t \geq 0$, used to establish the final value.

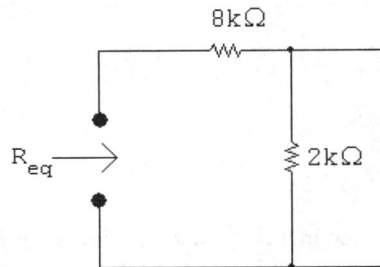

Figure 7.8: The circuit for Example 7.2, for $t \geq 0$, used to establish the time constant, τ.

We can make series and parallel combinations of the remaining resistors to find the Thévenin equivalent resistance we need for calculating the time constant. As seen in Fig. 7.8, the resistors are in series, so

$$R_{\text{Th}} = 2000 + 8000 = 10\text{k}\Omega.$$

Using this resistance, we can calculate the time constant as follows:

$$\tau = R_{\text{Th}}C = (10{,}000)(50 \times 10^{-9}) = 0.5 \text{ ms}.$$

4. Now that we have calculated the initial value of the capacitor voltage, the final value of the capacitor voltage, and the time constant, we can use these values to determine the voltage drop across the capacitor for $t \geq 0$:

$$v_C(t) = V_f + (V_o - V_f)e^{-t/\tau} = 50 + (100 - 50)e^{-t/(0.0005)}$$

$$= 50 + 50e^{-2000t} \text{ V}, \quad t \geq 0$$

5. The quantity requested in the original circuit shown in Fig. 7.5 is the current in the 2kΩ resistor, $i_R(t)$. To do this, we draw the circuit one final time for $t \geq 0$ with all components intact, as shown in Fig. 7.9.

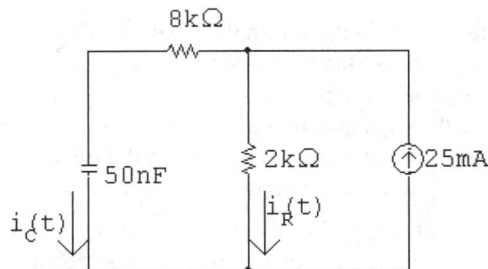

Figure 7.9: The circuit for Example 7.2, for $t \geq 0$.

We can find this current by writing a KCL equation at the top essential node:

$$i_R(t) = 25 \text{ mA} - i_C(t).$$

We need the current in the capacitor, $i_C(t)$, which we can get from the relationship between voltage and current in a capacitor:

$$i_C(t) = C\frac{dv_C(t)}{dt} = (50 \times 10^{-9})(-2000)(50e^{-2000t}) = -5e^{-2000t} \text{ mA}$$

Thus,

$$i_R(t) = 25 - (-5e^{-2000t}) = 25 + 5e^{-2000t} \text{mA}, \quad t \geq 0^+$$

Now try using the first-order circuit analysis method for each of the practice problems below.

Practice Problem 7.1

Find $v_C(t)$ for the circuit in Fig. 7.10.

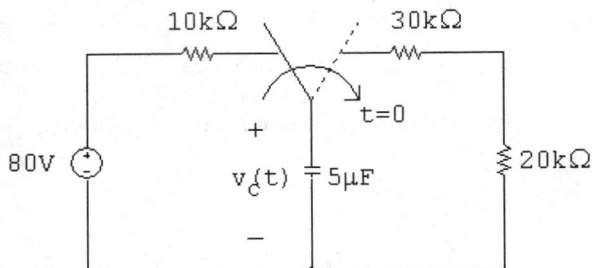

Figure 7.10: The circuit for Practice Problem 7.1.

1. Draw the circuit in Fig. 7.10 for $t < 0$, replacing the capacitor with an open circuit whose voltage is labeled V_o. Find the value of V_o, which is the initial capacitor voltage.

2. Draw the circuit in Fig. 7.10 for $t \geq 0$, replacing the capacitor with an open circuit whose voltage is labeled V_f. Find the value of V_f, which is the final capacitor voltage.

3. Draw the circuit in Fig. 7.10 for $t \geq 0$, replacing the capacitor with an open circuit. Find the value of the equivalent resistance seen from the open circuit where the capacitor was, which is the Thévenin equivalent resistance, R_{Th}. Then use this equivalent resistance to find the time constant $\tau = R_{Th}C$.

4. Find the expression for the voltage drop across the capacitor, $v_C(t)$ from the initial voltage, the final voltage and the time constant.

5. Since there are no other voltages or currents requested in the circuit in Fig. 7.10, the analysis is complete.

Practice Problem 7.2

Find $v_R(t)$ for the circuit in Fig. 7.11.

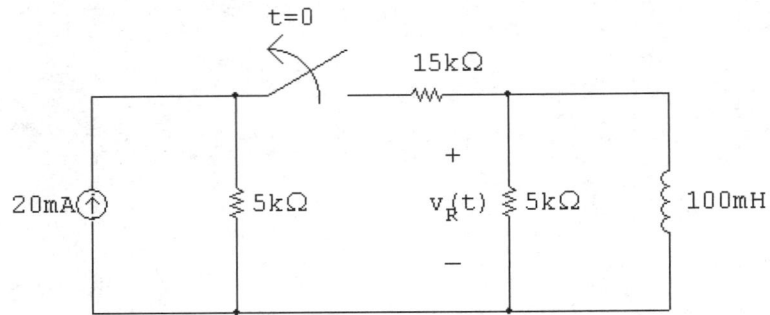

Figure 7.11: The circuit for Practice Problem 7.2.

1. Draw the circuit in Fig. 7.11 for $t < 0$, replacing the inductor with a short circuit whose current is labeled I_o. Find the value of I_o, which is the initial inductor current.

2. Draw the circuit in Fig. 7.11 for $t \geq 0$, replacing the inductor with a short circuit whose current is labeled I_f. Find the value of I_f, which is the final inductor current.

3. Draw the circuit in Fig. 7.11 for $t \geq 0$, replacing the inductor with an open circuit. Find the value of the equivalent resistance seen from the open circuit where the inductor was, which is the Thévenin equivalent resistance, R_{Th}. Then use this equivalent resistance to find the time constant $\tau = L/R_{Th}$.

4. Find the expression for the current in the inductor, $i_L(t)$ from the initial current, the final current and the time constant.

5. Draw the circuit in Fig. 7.11 for $t \geq 0$ with all components intact. Using this circuit, the expression for the inductor current $i_L(t)$, and circuit analysis, find the requested quantity $v_R(t)$.

Practice Problem 7.3

Find $v_C(t)$ for the circuit in Fig. 7.12.

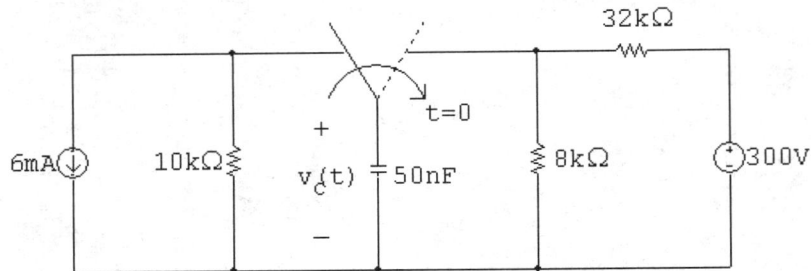

Figure 7.12: The circuit for Practice Problem 7.3.

1. Draw the circuit in Fig. 7.12 for $t < 0$, replacing the capacitor with an open circuit whose voltage is labeled V_o. Find the value of V_o, which is the initial capacitor voltage.

2. Draw the circuit in Fig. 7.12 for $t \geq 0$, replacing the capacitor with an open circuit whose voltage is labeled V_f. Find the value of V_f, which is the final capacitor voltage.

3. Draw the circuit in Fig. 7.12 for $t \geq 0$, replacing the capacitor with an open circuit. Find the value of the equivalent resistance seen from the open circuit where the capacitor was, which is the Thévenin equivalent resistance, R_{Th}. Then use this equivalent resistance to find the time constant $\tau = R_{Th}C$.

4. Find the expression for the voltage drop across the capacitor, $v_C(t)$ from the initial voltage, the final voltage and the time constant.

5. Since there are no other voltages or currents requested in the circuit in Fig. 7.12, the analysis is complete.

Practice Problem 7.4

Find $i_R(t)$ for the circuit in Fig. 7.13.

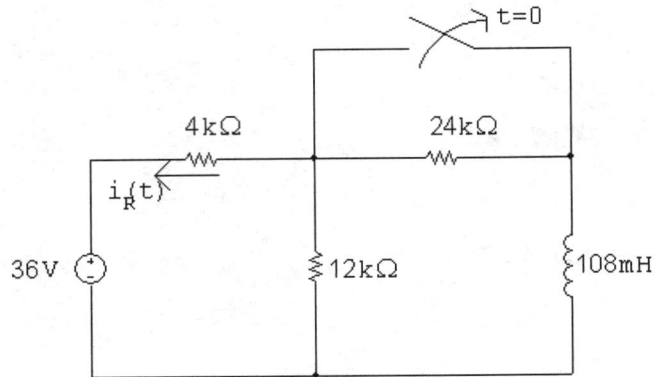

Figure 7.13: The circuit for Practice Problem 7.4.

1. Draw the circuit in Fig. 7.13 for $t < 0$, replacing the inductor with a short circuit whose current is labeled I_o. Find the value of I_o, which is the initial inductor current.

2. Draw the circuit in Fig. 7.13 for $t \geq 0$, replacing the inductor with a short circuit whose current is labeled I_f. Find the value of I_f, which is the final inductor current.

3. Draw the circuit in Fig. 7.13 for $t \geq 0$, replacing the inductor with an open circuit. Find the value of the equivalent resistance seen from the open circuit where the inductor was, which is the Thévenin equivalent resistance, R_{Th}. Then use this equivalent resistance to find the time constant $\tau = L/R_{Th}$.

4. Find the expression for the current in the inductor, $i_L(t)$ from the initial current, the final current and the time constant.

5. Draw the circuit in Fig. 7.13 for $t \geq 0$ with all components intact. Using this circuit, the expression for the inductor current $i_L(t)$, and circuit analysis, find the requested quantity $i_R(t)$.

Practice Problem 7.5

Find $i_R(t)$ for the circuit in Fig. 7.14.

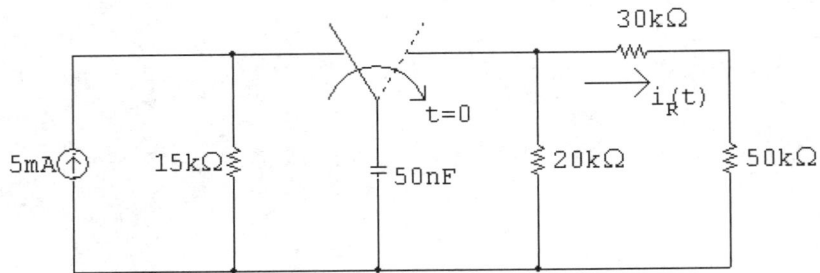

Figure 7.14: The circuit for Practice Problem 7.5.

1. Draw the circuit in Fig. 7.14 for $t < 0$, replacing the capacitor with an open circuit whose voltage is labeled V_o. Find the value of V_o, which is the initial capacitor voltage.

2. Draw the circuit in Fig. 7.14 for $t \geq 0$, replacing the capacitor with an open circuit whose voltage is labeled V_f. Find the value of V_f, which is the final capacitor voltage.

3. Draw the circuit in Fig. 7.14 for $t \geq 0$, replacing the capacitor with an open circuit. Find the value of the equivalent resistance seen from the open circuit where the capacitor was, which is the Thévenin equivalent resistance, R_{Th}. Then use this equivalent resistance to find the time constant $\tau = R_{Th}C$.

4. Find the expression for the voltage drop across the capacitor, $v_C(t)$ from the initial voltage, the final voltage and the time constant.

5. Draw the circuit in Fig. 7.14 for $t \geq 0$ with all components intact. Using this circuit, the expression for the capacitor voltage $v_C(t)$, and circuit analysis, find the requested quantity $i_R(t)$.

Practice Problem 7.6

Find $v_R(t)$ for the circuit in Fig. 7.15.

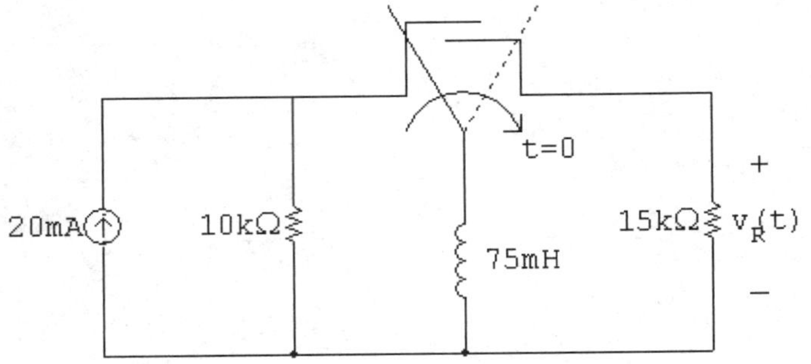

Figure 7.15: The circuit for Practice Problem 7.6.

1. Draw the circuit in Fig. 7.15 for $t < 0$, replacing the inductor with a short circuit whose current is labeled I_o. Find the value of I_o, which is the initial inductor current.

2. Draw the circuit in Fig. 7.15 for $t \geq 0$, replacing the inductor with a short circuit whose current is labeled I_f. Find the value of I_f, which is the final inductor current.

3. Draw the circuit in Fig. 7.15 for $t \geq 0$, replacing the inductor with an open circuit. Find the value of the equivalent resistance seen from the open circuit where the inductor was, which is the Thévenin equivalent resistance, R_{Th}. Then use this equivalent resistance to find the time constant $\tau = L/R_{Th}$.

4. Find the expression for the current in the inductor, $i_L(t)$ from the initial current, the final current and the time constant.

5. Draw the circuit in Fig. 7.15 for $t \geq 0$ with all components intact. Using this circuit, the expression for the inductor current $i_L(t)$, and circuit analysis, find the requested quantity $v_R(t)$.

Practice Problem 7.7

Find $v_R(t)$ for the circuit in Fig. 7.16.

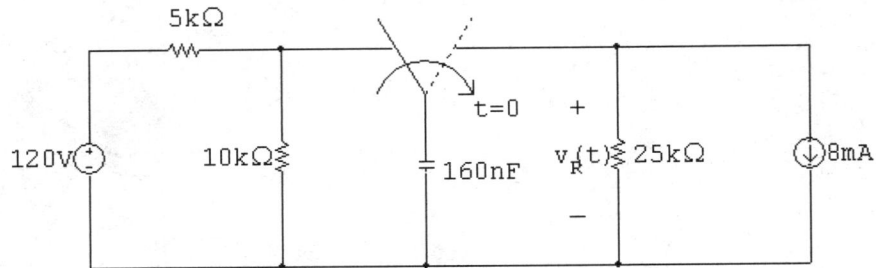

Figure 7.16: The circuit for Practice Problem 7.7.

1. Draw the circuit in Fig. 7.16 for $t < 0$, replacing the capacitor with an open circuit whose voltage is labeled V_o. Find the value of V_o, which is the initial capacitor voltage.

2. Draw the circuit in Fig. 7.16 for $t \geq 0$, replacing the capacitor with an open circuit whose voltage is labeled V_f. Find the value of V_f, which is the final capacitor voltage.

3. Draw the circuit in Fig. 7.16 for $t \geq 0$, replacing the capacitor with an open circuit. Find the value of the equivalent resistance seen from the open circuit where the capacitor was, which is the Thévenin equivalent resistance, R_{Th}. Then use this equivalent resistance to find the time constant $\tau = R_{Th}C$.

4. Find the expression for the voltage drop across the capacitor, $v_C(t)$ from the initial voltage, the final voltage and the time constant.

5. Draw the circuit in Fig. 7.16 for $t \geq 0$ with all components intact. Using this circuit, the expression for the capacitor voltage $v_C(t)$, and circuit analysis, find the requested quantity $v_R(t)$.

Practice Problem 7.8

Find $i_L(t)$ for the circuit in Fig. 7.17.

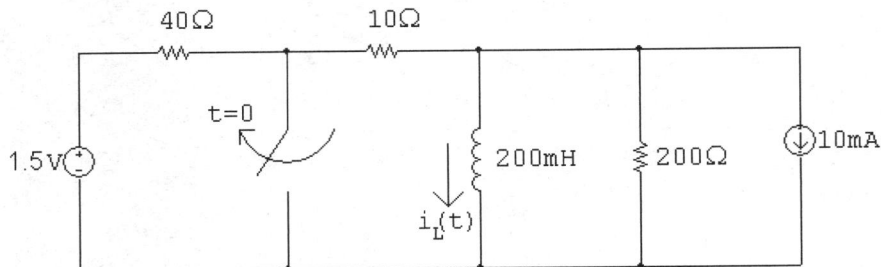

Figure 7.17: The circuit for Practice Problem 7.8.

1. Draw the circuit in Fig. 7.17 for $t < 0$, replacing the inductor with a short circuit whose current is labeled I_o. Find the value of I_o, which is the initial inductor current.

2. Draw the circuit in Fig. 7.17 for $t \geq 0$, replacing the inductor with a short circuit whose current is labeled I_f. Find the value of I_f, which is the final inductor current.

3. Draw the circuit in Fig. 7.17 for $t \geq 0$, replacing the inductor with an open circuit. Find the value of the equivalent resistance seen from the open circuit where the inductor was, which is the Thévenin equivalent resistance, R_{Th}. Then use this equivalent resistance to find the time constant $\tau = L/R_{Th}$.

4. Find the expression for the current in the inductor, $i_L(t)$ from the initial current, the final current and the time constant.

5. Since there are no other voltages or currents requested in the circuit in Fig. 7.17, the analysis is complete.

Reading

- in *Electric Circuits*, ninth edition:
    - ♦ Section 6.1 — the inductor
    - ♦ Section 6.2 — the capacitor
    - ♦ Section 7.1 — natural response of RL circuits
    - ♦ Section 7.2 — natural response of RC circuits
    - ♦ Section 7.3 — step response of RL and RC circuits
    - ♦ Section 7.4 — general solution for natural and step response

- Workbook section — Combining Resistors in Series and in Parallel

- Workbook section — Thévenin and Norton Equivalents

Additional Problems

- 7.6
- 7.12
- 7.14
- 7.16
- 7.23
- 7.27
- 7.30
- 7.35 – 7.37
- 7.41 – 7.42
- 7.50 – 7.60
- 7.63

Solutions

- Practice Problem 7.1:

$$V_o = 80 \text{ V} \qquad V_f = 0 \text{ V} \qquad \tau = 0.25 \text{ ms} \qquad v_C(t) = 80e^{-4t} \text{ V}, \quad t \geq 0$$

- Practice Problem 7.2:

$$I_o = 5 \text{ mA} \qquad I_f = 0 \text{ A} \qquad \tau = 20\mu\text{s} \qquad i_R(t) = -5e^{-50,000t} \text{ A}, \quad t \geq 0^+$$

- Practice Problem 7.3:

$$V_o = -60 \text{ V} \qquad V_f = 60 \text{ V} \qquad \tau = 320\mu\text{s} \qquad v_C(t) = 60 - 120e^{-3125t} \text{ V}, \quad t \geq 0$$

- Practice Problem 7.4:

$$I_o = 9 \text{ mA} \qquad I_f = 1 \text{ mA} \qquad \tau = 4\mu\text{s} \qquad i_R(t) = -3 - 6e^{-250,000t} \text{ mA}, \quad t \geq 0^+$$

- Practice Problem 7.4:

$$I_o = 9 \text{ mA} \qquad I_f = 1 \text{ mA} \qquad \tau = 4\mu s \qquad i_R(t) = -3 - 6e^{-250,000t} \text{ mA}, \quad t \geq 0^+$$

- Practice Problem 7.5:

$$V_o = 75 \text{ V} \qquad V_f = 0 \text{ V} \qquad \tau = 800\mu s \qquad i_R(t) = 0.9375e^{-1250t} \text{ mA}, \quad t \geq 0^+$$

- Practice Problem 7.6:

$$I_o = 20 \text{ mA} \qquad I_f = 0 \text{ A} \qquad \tau = 5\mu s \qquad v_R(t) = -300e^{-200,000t} \text{ V}, \quad t \geq 0^+$$

- Practice Problem 7.7:

$$V_o = 80 \text{ V} \qquad V_f = -200 \text{ V} \qquad \tau = 4 \text{ ms} \qquad v_R(t) = -200 + 280e^{-250t} \text{ V}, \quad t \geq 0^+$$

- Practice Problem 7.8:

$$I_o = -10 \text{ mA} \qquad I_f = 20 \text{ mA} \qquad \tau = 5 \text{ ms} \qquad i_R(t) = 20 - 30e^{-200t} \text{ mA}, \quad t \geq 0$$

Chapter 8

Natural and Step Response of Second-Order (RLC) Circuits

Here we review and then practice the techniques that enable us to analyze a limited group of circuits. These are circuits containing one equivalent resistor, one equivalent inductor, and one equivalent capacitor. The resistor, inductor, and capacitor can be connect in series or in parallel. Both the inductor and the capacitor may have initial stored energy. The use of the phrase "one equivalent" means that if the circuit contains two or more resistors, for example, they must be arranged in such a way that they can be combined in series and in parallel to form one single equivalent resistor. The same holds for circuits that contain two or more inductors, or two or more capacitors. These circuits are referred to as RLC circuits, and are also called **second-order circuits**, because their describing equation is a second-order differential equation.

These circuits usually contain a switch that is in one position for $t < 0$, switches positions at $t = 0$, and remains at that second position indefinitely. When the switch is in its first position, there may be an independent current or voltage source in the circuit as well, used to generate the energy that the inductor or capacitor will have stored at $t = 0$. When the switch moves to its second position, there may or may not be an independent current or voltage source in the circuit. If there is, it continues to supply energy to the circuit indefinitely, and we call the analysis a **step response** problem. If there is not an independent source in the circuit for $t \geq 0$, then the energy initially stored is dissipated to the resistor and we call the analysis a **natural response** problem. Fortunately the natural response problem and the step response problem are closely related, so we can use the same circuit analysis technique for both problems.

Analyzing RLC circuits connected in series is very similar to analyzing RLC circuits connected in parallel so we can also use the same circuit analysis technique for both circuits. There are basically five steps in the analysis: find the initial conditions, which consist of the initial current in the inductor and the initial voltage drop across the capacitor; find the final values, which are the final current in the inductor and the final voltage drop across the capacitor; find the **neper frequency**, α, which equals $1/2RC$ for the parallel circuit and equals $R/2L$ for the series circuit and the **resonant radian frequency**, ω_o, which is $\sqrt{1/LC}$, compare α^2 and ω_o^2 to determine whether the response type is **overdamped**, **underdamped**, or **critically damped** and write down the form of the response; use the response form to determine the value of the response at $t = 0$ and the value of the first derivative of the response at $t = 0$, then use the circuit to determine the same two quantities, providing enough information to solve for the unknown coefficients in the response form; and finally, use the calculated response to determine the values of any other requested voltages and currents in the circuit. For the natural response of the parallel RLC circuit the response we calculate is the voltage drop across the parallel elements. For the natural response of the series RLC circuit the response we calculate is the current through the series elements. For the step response problems, we will calculate the only quantity that has a non-zero final value — that is the voltage drop across the capacitor in the series RLC circuit and the current through the inductor in the parallel RLC circuit.

The analysis method for RLC circuits can be broken into the following steps:

1. Redraw the circuit as it appears for $t < 0$, replacing the switch with an open circuit if it is open, and with a short circuit if it is closed. Since it is assumed that the switch has been in this position for a long time, this places any inductors and capacitors in the presence of a constant source. Therefore, an inductor should be replaced by a short circuit and a capacitor should be replaced by an open circuit. Using this circuit for $t < 0$, calculate the current through the short circuit, which is the initial current I_o and the voltage drop across the open circuit, which is the initial voltage drop V_o.

2. Redraw the circuit as it appears for $t \geq 0$, replacing the switch with an open circuit if it is open and with a short circuit if it is closed. If there are no independent sources in the circuit, this is the natural response problem. The final value of the

voltage drop across the capacitor $V_f = 0$ and the final value of the current through the inductor $I_f = 0$, since all of the initially stored energy in the inductor and capacitor will be dissipated by the resistor as $t \to \infty$.

If there is an independent source in the circuit for $t \geq 0$ this is the step response problem. The inductor and capacitor will have been in the presence of this independent source for a long time as $t \to \infty$ so replace the inductor with a short circuit and the capacitor with an open circuit. Calculate the current in the short circuit, which is the final current, I_f, and the voltage drop across the open circuit, which is the final voltage V_f. Make sure the direction of the current arrow is the same when computing I_o in Step 1 and when computing I_f in this step. Make sure that the polarity of the voltage when you are calculating the initial voltage V_o is the same as the polarity of the voltage when you are calculating the final value V_f.

3. Determine the response type by calculating ω_o and α. For both series and parallel RLC circuits,

$$\omega_o = \sqrt{\frac{1}{LC}}$$

The computation of α depends on the configuration of the circuit:

$$\text{For series-connected } RLC \text{ circuits} \quad \alpha = \frac{R}{2L};$$

$$\text{For parallel-connected } RLC \text{ circuits} \quad \alpha = \frac{1}{2RC}$$

Then compare α^2 and ω_o^2 to determine the form of the response:

- If $\alpha^2 > \omega_o^2$, the response type is overdamped and of the form $X_f + A_1 e^{-s_1 t} + A_2 e^{-s_2 t}$.
- If $\alpha^2 < \omega_o^2$, the response type is underdamped and of the form $X_f + (B_1 \cos \omega_d t + B_2 \sin \omega_d t)e^{-\alpha t}$.
- If $\alpha^2 = \omega_o^2$, the response type is critically damped and of the form $X_f + D_1 t e^{-\alpha t} + D_2 e^{-\alpha t}$.

In the above equations, X_f is the final value of the voltage or the current, depending on whether you are determining the response form for a voltage or a current. For the natural response problem $X_f = 0$ and you will determine the voltage drop for the parallel RLC circuit and the current for the series RLC circuit. For the step response problem, the non-zero final value will be the inductor current for the parallel RLC circuit and the capacitor voltage drop form the series RLC circuit.

4. Write the equation describing the response you are calculating. If the response form is overdamped, you will need to calculate s_1 and s_2 from the equation
$$s_{1,2} = -\alpha \pm \sqrt{\alpha^2 - \omega_o^2}$$

If the response form is underdamped, you will need to calculate ω_d from the equation

$$\omega_d = \sqrt{\omega_o^2 - \alpha^2}$$

If the response form is critically damped, you need make no additional calculations. Regardless of the response form you will have two unspecified coefficients whose values will be used to satisfy the initial conditions. Evaluate the initial value of the response (at $t = 0$) and the initial value of the first derivative of the response. These equations will involve the unknown coefficients. Then use the circuit to determine the initial value of the response, which will be I_o or V_o determined in Step 1, and the initial value of the first derivative of the response, whose value will also involve I_o or V_o. Then equate the initial values from the equation and its first derivative with the initial values from the circuit quantities. This provides two equations which, when solved simultaneously, will yield the values of the unknown coefficients. Complete this step by writing the response using the values of the unknown coefficients.

5. If a voltage or current other than the one you calculated in Step 4 was requested for this circuit, use the calculated value to determine the requested value. If the current or voltage you calculated was the one sought for the circuit, you are finished.

The following three examples illustrate the process of analyzing second-order circuits. There is one example for each of the three different response types. The examples illustrate both the series RLC circuit and the parallel RLC circuit and also illustrate both a natural response problem and a step response problem.

Example 8.1

Find $v_R(t)$ for the circuit in Fig. 8.1 for $t \geq 0$.

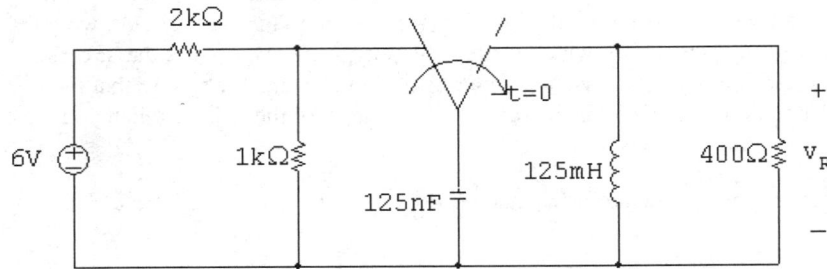

Figure 8.1: The circuit for Example 8.1

Solution

1. Redraw the circuit in Fig. 8.1 with the switch in its left hand position. This is the circuit for $t < 0$ and is used to establish the initial conditions. It is assumed that the switch has been in this position for a long time, so the inductor is replaced with a short circuit with current I_o and the capacitor is replaced by an open circuit with voltage drop V_o. The resulting circuit is shown in Fig. 8.2.

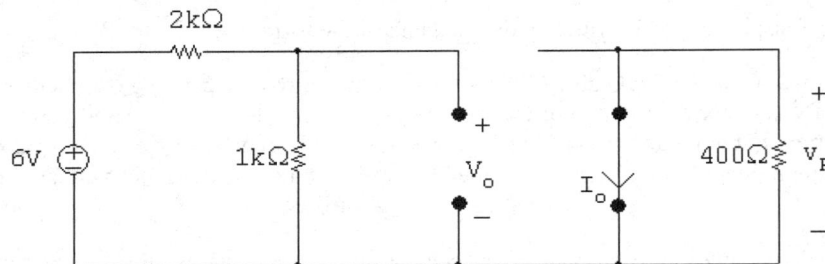

Figure 8.2: The circuit for Example 8.1, for $t < 0$, used to establish the initial conditions.

We must analyze this circuit to find I_o and V_o. I_o is easy because there is no current flowing in the right hand side of the circuit, due to the position of the switch. Therefore,

$$I_o = 0 \text{ A.}$$

To find V_o we use voltage division as follows:

$$V_o = \frac{1000}{1000 + 2000}(6) = 2 \text{ V.}$$

2. Redraw the circuit in Fig. 8.1 with the switch in its right hand position. This is the circuit for $t \geq 0$ and is used to establish the final values, since the circuit will be in this configuration as $t \to \infty$. It is assumed that the switch has been in this position for a long time, so the inductor is replaced with a short circuit with current I_f and the capacitor is replaced with an open circuit with voltage drop V_f. The resulting circuit is shown in Fig. 8.3.

As you can see, there are no independent sources in this circuit, so the stored energy in the capacitor will dissipate in the resistor leaving no energy in the capacitor. There was never any stored energy in the inductor. Thus, the final voltage $V_f = 0$V and the final current $I_f = 0$A.

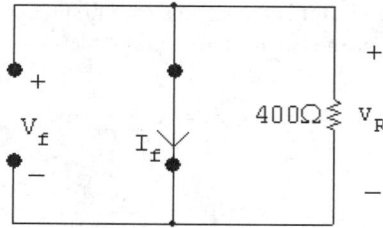

Figure 8.3: The circuit for Example 8.1, for $t \geq 0$, used to establish the final values.

3. To find α and ω_o we consider the values of the resistor, the inductor, and the capacitor in the circuit for $t \geq 0$, as shown in Fig. 8.4.

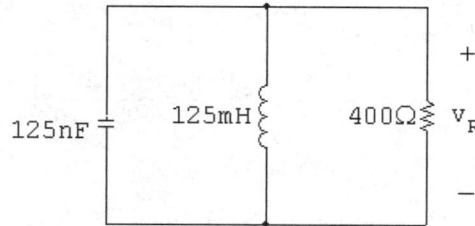

Figure 8.4: The circuit for Example 8.1, for $t \geq 0$, used to calculate α and ω_o.

We substitute these values into the equations appropriate for the parallel RLC circuit:

$$\alpha = \frac{1}{2RC} = \frac{1}{2(400)(125 \times 10^{-9})} = 10,000 \text{ rad/sec};$$

$$\omega_o = \sqrt{\frac{1}{LC}} = \sqrt{\frac{1}{(0.125)(125 \times 10^{-9})}} = 8000 \text{ rad/sec}.$$

Now we compare the values of α^2 and ω_O^2 to determine the response type. Since $\alpha^2 > \omega_o^2$, the response is overdamped, and since this is the parallel RLC natural response problem, the response we determine is the voltage across the parallel components, $v_R(t)$.

4. In order to specify an overdamped response, we need to calculate the values of the complex frequencies s_1 and s_2:

$$s_{1,2} = -\alpha \pm \sqrt{\alpha^2 - \omega_o^2} = -10,000 \pm \sqrt{10,000^2 - 8000^2} = -10,000 \pm 6000$$

Thus,

$$s_1 = -4000 \text{ rad/sec} \quad \text{and} \quad s_2 = -16,000 \text{ rad/sec}.$$

Therefore, the response is

$$v_R(t) = A_1 e^{-4000t} + A_2 e^{-16,000t} \text{ V}.$$

To calculate the coefficients A_1 and A_2, we need two equations. The first equation is the result of evaluating the response $v_R(t)$ at $t = 0$ and settling the result equal to the initial value of the voltage from the circuit, V_o:

$$v_R(0) = A_1 + A_2 = V_o = 6 \text{ V}.$$

The second equation is the result of evaluating the first derivative of the response $v_R(t)$ at $t = 0$ and setting the result equal to the initial value of the first derivative of the voltage from the circuit. The first derivative of the response $v_R(t)$ at $t = 0$ is

$$\frac{dv_R(0)}{dt} = -4000A_1 - 16,000A_2$$

The first derivative of the $v_R(t)$ from the circuit is the same as the first derivative of the voltage across the capacitor, since the circuit components are in parallel. For the capacitor we know that

$$i_C(t) = C\frac{dv_C(t)}{dt}$$

so

$$i_C(0) = C\frac{dv_C(0)}{dt}$$

and

$$\frac{dv_C(0)}{dt} = \frac{1}{C}i_C(0).$$

We don't know the value of the initial current in the capacitor, $i_C(0)$, but we do know that the sum of the capacitor current, the inductor current, and the resistor current must be zero, from KCL. Therefore,

$$\frac{dv_C(0)}{dt} = \frac{1}{C}i_C(0) = \frac{1}{C}[-i_L(0) - i_R(0)]$$

$$= \frac{1}{C}\left[-I_o - \frac{V_o}{R}\right] = \frac{1}{125 \times 10^{-9}}\left[0 - \frac{2}{400}\right]$$

$$= -40,000 \text{ V/s}.$$

To summarize, the two equations used to solve for A_1 and A_2 are

$$A_1 + A_2 = 2$$
$$-4000A_1 - 16,000A_2 = -40,000$$

Since these equations are already in standard form, we can use the calculator to solve them:

$$A_1 = -0.667; \qquad A_2 = 2.667.$$

Thus,

$$v_R(t) = -0.667e^{-4000t} + 2.667e^{-16,000t} \text{ V}, \quad t \geq 0.$$

5. Since the voltage drop across the parallel components was the only quantity requested in the original circuit shown in Fig. 8.1, no further analysis is required.

Example 8.2

Find $v_C(t)$ for the circuit in Fig. 8.5 for $t \geq 0$.

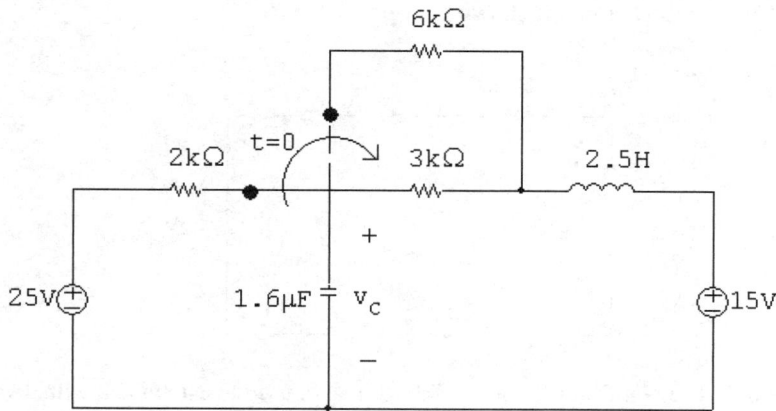

Figure 8.5: The circuit for Example 8.2

Solution

1. Redraw the circuit in Fig. 8.5 with the switch in its down position. This is the circuit for $t < 0$ and is used to establish the initial conditions. It is assumed that the switch has been in this position for a long time, so the inductor is replaced with a short circuit with current I_o and the capacitor is replaced by an open circuit with voltage drop V_o. The resulting circuit is shown in Fig. 8.6.

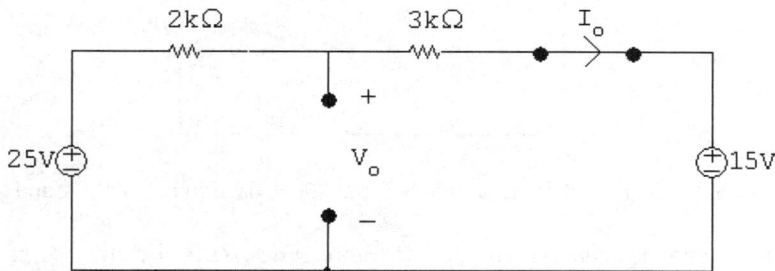

Figure 8.6: The circuit for Example 8.2, for $t < 0$, used to establish the initial conditions.

We must analyze this circuit to find I_o and V_o. I_o can be found by applying Ohm's law. Therefore,

$$I_o = \frac{25 - 15}{2000 + 3000} = 5 \text{ mA}.$$

To find V_o we write a node voltage equation as follows:

$$\frac{V_o - 25}{2000} + \frac{V_o - 15}{3000} = 0.$$

Solving for V_o we get

$$V_o = 21 \text{ V}.$$

2. Redraw the circuit in Fig. 8.5 with the switch in its up position. This is the circuit for $t \geq 0$ and is used to establish the final values, since the circuit will be in this configuration as $t \to \infty$. It is assumed that the switch has been in this position for a long time, so the inductor is replaced with a short circuit with current I_f and the capacitor is replaced with an open circuit with voltage drop V_f. Notice that we have also combined the parallel resistors into a single resistor with the value $3000\|6000 = 2$ kΩ. The resulting circuit is shown in Fig. 8.7.

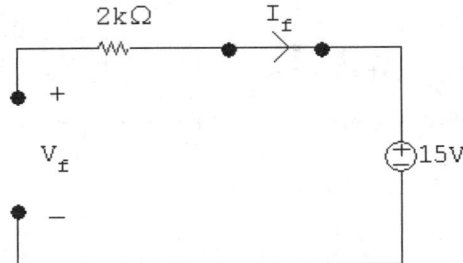

Figure 8.7: The circuit for Example 8.2, for $t \geq 0$, used to establish the final values.

As you can see, there is an independent source in this circuit, so this is a step response problem. Because of the open circuit created by the capacitor, there is no current, so

$$I_f = 0 \text{ A} \quad \text{and} \quad V_f = 15 \text{ V.}$$

3. To find α and ω_o we consider the values of the resistor, the inductor, and the capacitor in the circuit for $t \geq 0$, as shown in Fig. 8.8.

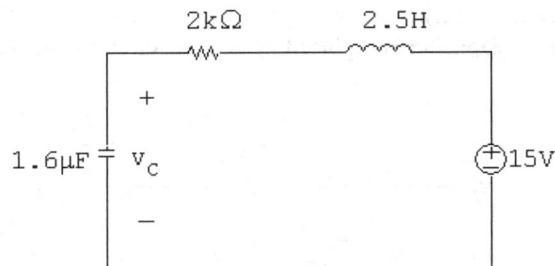

Figure 8.8: The circuit for Example 8.2, for $t \geq 0$, used to calculate α and ω_o.

We substitute these values into the equations appropriate for the series RLC circuit:

$$\alpha = \frac{R}{2L} = \frac{3000}{2(2.5)} = 400 \text{ rad/sec;}$$

$$\omega_o = \sqrt{\frac{1}{LC}} = \sqrt{\frac{1}{(2.5)(1.6 \times 10^{-6})}} = 500 \text{ rad/sec.}$$

Now we compare the values of α^2 and ω_o^2 to determine the response type. Since $\alpha^2 < \omega_o^2$, the response is underdamped, and since this is the series RLC step response problem and the only non-zero final value is the voltage drop across the capacitor, the response we determine is the voltage across the capacitor, $v_C(t)$.

4. In order to specify an underdamped response, we need to calculate the values of the damped radian frequency ω_d:

$$\omega_d = \sqrt{\omega_o^2 - \alpha^2} = \sqrt{500^2 - 400^2} = 300 \text{ rad/sec.}$$

Therefore, the response is

$$v_C(t) = V_f + (B_1 \cos 300t + B_2 \sin 300t)e^{-400t} \text{ V.}$$

To calculate the coefficients B_1 and B_2, we need two equations. The first equation is the result of evaluating the response $v_C(t)$ at $t = 0$ and settling the result equal to the initial value of the voltage from the circuit, V_o:

$$v_C(0) = V_f + B_1 = V_o = 21 \text{ V}.$$

The second equation is the result of evaluating the first derivative of the response $v_C(t)$ at $t = 0$ and setting the result equal to the initial value of the first derivative of the voltage from the circuit. The first derivative of $v_C(t)$ at $t = 0$ is

$$\frac{dv_C(0)}{dt} = -400B_1 + 500B_2$$

The first derivative of the $v_C(t)$ from the circuit can be found from the describing equation for the voltage and current in a capacitor:

$$i_C(t) = C\frac{dv_C(t)}{dt}$$

so

$$i_C(0) = C\frac{dv_C(0)}{dt}$$

and

$$\frac{dv_C(0)}{dt} = \frac{1}{C}i_C(0).$$

The initial current in the capacitor, $i_C(0)$, has the same value as the initial value of the current in the inductor, since they are in series, but the opposite sign. Therefore,

$$\frac{dv_C(0)}{dt} = \frac{1}{C}i_C(0) = \frac{1}{C}I_o = \frac{1}{1.6 \times 10^{-6}}(-0.005) = -3125.$$

To summarize, the two equations used to solve for A_1 and A_2 are

$$\begin{aligned} B_1 &= 21 - 15 = 6 \\ -400B_1 + 300B_2 &= -3125 \end{aligned}$$

Since these equations are already in standard form, we can use the calculator to solve them:

$$B_1 = 6; \qquad B_2 = -2.417.$$

Thus,

$$v_C(t) = 15 + (6\cos 300t - 2.417\sin 300t)e^{-400t} \text{ V}, \quad t \geq 0.$$

5. Since the voltage drop across the capacitor was the only quantity requested in the original circuit shown in Fig. 8.5, no further analysis is required.

Example 8.3

There is no initial energy stored in the circuit in Fig. 8.9. Find $i_L(t)$ for this circuit for $t \geq 0$.

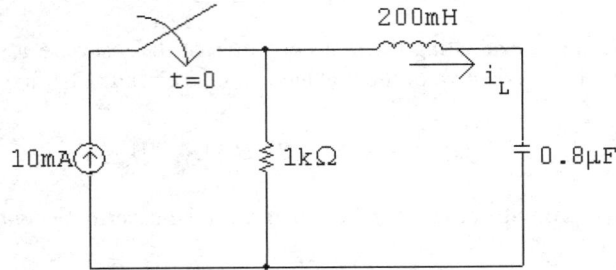

Figure 8.9: The circuit for Example 8.3

Solution

1. Since we have already been told that there is no initial stored energy in the circuit, we don't need to analyze the circuit for $t < 0$ to find the initial conditions, since they are both zero. Therefore,

$$I_o = 0 \text{ A} \qquad \text{and} \qquad V_o = 0 \text{ V}.$$

2. Redraw the circuit in Fig. 8.9 with the switch closed. This is the circuit for $t \geq 0$ and is used to establish the final values, since the circuit will be in this configuration as $t \to \infty$. It is assumed that the switch has been in this position for a long time, so the inductor is replaced with a short circuit with current I_f and the capacitor is replaced with an open circuit with voltage drop V_f. We have also performed a source transformation to turn the parallel combination of the current source and resistor into a series combination of a voltage source and the same resistor. The resulting circuit is shown in Fig. 8.10.

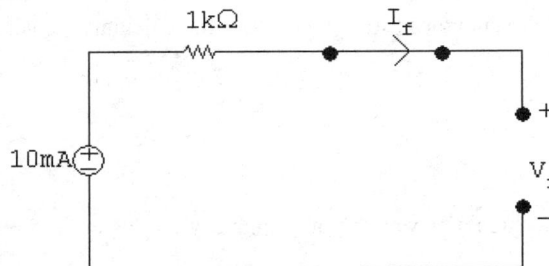

Figure 8.10: The circuit for Example 8.3, for $t \geq 0$, used to establish the final values.

As you can see, there is an independent source in this circuit, so this is a step response problem. Because of the open circuit created by the capacitor, there is no current, so

$$I_f = 0 \text{ A} \qquad \text{and} \qquad V_f = 10 \text{ V}.$$

3. To find α and ω_o we consider the values of the resistor, the inductor, and the capacitor in the circuit for $t \geq 0$, as shown in Fig. 8.11.

We substitute these values into the equations appropriate for the series RLC circuit:

$$\alpha = \frac{R}{2L} = \frac{1000}{2(0.2)} = 2500 \text{ rad/sec};$$

$$\omega_o = \sqrt{\frac{1}{LC}} = \sqrt{\frac{1}{(0.2)(0.8 \times 10^{-6})}} = 2500 \text{ rad/sec}.$$

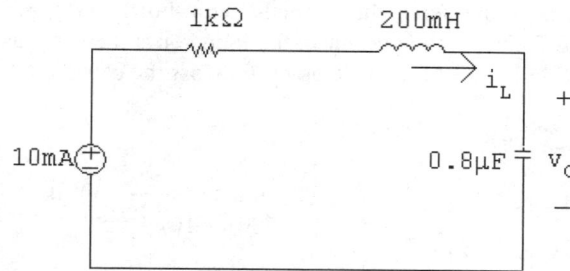

Figure 8.11: The circuit for Example 8.3, for $t \geq 0$, used to calculate α and ω_o.

Now we compare the values of α^2 and ω_O^2 to determine the response type. Since $\alpha^2 = \omega_o^2$, the response is critically damped, and since this is the series RLC step response problem and the only non-zero final value is the voltage drop across the capacitor, the response we determine is the voltage across the capacitor, $v_C(t)$.

4. We don't need to make any additional calculations for the critically damped response type. Therefore, the response is

$$v_C(t) = V_f + D_1 t e^{-2500t} + D_2 e^{-2500t} \text{ V.}$$

To calculate the coefficients D_1 and D_2, we need two equations. The first equation is the result of evaluating the response $v_C(t)$ at $t = 0$ and settling the result equal to the initial value of the voltage from the circuit, V_o:

$$v_C(0) = V_f + D_2 = V_o = 0 \text{ V.}$$

The second equation is the result of evaluating the first derivative of the response $v_C(t)$ at $t = 0$ and setting the result equal to the initial value of the first derivative of the voltage from the circuit. The first derivative of the $v_C(t)$ at $t = 0$ is

$$\frac{dv_C(0)}{dt} = D_1 - 2500 D_2$$

The first derivative of $v_C(t)$ from the circuit can be found from the describing equation for the voltage and current in a capacitor:

$$i_C(t) = C \frac{dv_C(t)}{dt}$$

so

$$i_C(0) = C \frac{dv_C(0)}{dt}$$

and

$$\frac{dv_C(0)}{dt} = \frac{1}{C} i_C(0).$$

The initial current in the capacitor, $i_C(0)$, has the same value as the initial value of the current in the inductor, since they are in series. Therefore,

$$\frac{dv_C(0)}{dt} = \frac{1}{C} i_C(0) = \frac{1}{C} I_o = \frac{1}{1.6 \times 10^{-6}} (0) = 0.$$

To summarize, the two equations used to solve for D_1 and D_2 are

$$\begin{aligned} D_2 &= 0 - 10 = -10 \\ D_1 - 2500 D_2 &= 0 \end{aligned}$$

Since these equations are already in standard form, we can use the calculator to solve them:

$$D_1 = -25,000; \qquad D_2 = -10.$$

Thus,

$$v_C(t) = 10 - 25,000 t e^{-2500t} - 10 e^{-2500t} \text{ V,} \quad t \geq 0.$$

5. Now we must calculate the quantity requested in the original circuit shown in Fig. 8.5, which is the current in the inductor, $i_L(t)$. Since this is a series RLC circuit, the current in the inductor is the same as the current in the capacitor, whose value we can calculate from the derivative of the voltage drop across the capacitor:

$$
\begin{aligned}
i_L(t) &= i_C(t) = C\frac{dv(t)}{dt} \\
&= (0.8 \times 10^{-6})[-25{,}000e - 2500t + (-2500)(-25{,}000)te^{-2500t} \\
&\qquad\qquad +(-2500)(-10)e^{-2500t}] \\
&= 50te^{-2500t} \text{ A}, \quad t \geq 0.
\end{aligned}
$$

Now try using the second-order circuit analysis method for each of the practice problems below.

Practice Problem 8.1

There is no initial stored energy in the circuit shown in Fig. 8.12. Find $v_C(t)$ for this circuit.

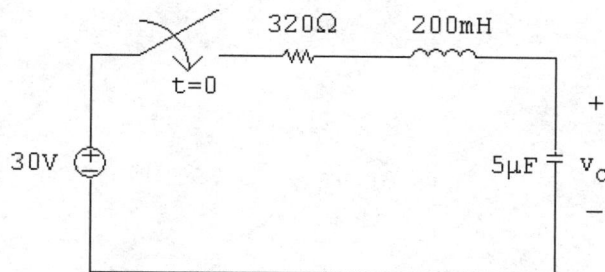

Figure 8.12: The circuit for Practice Problem 8.1.

1. Find the initial current through the inductor, I_o and the initial voltage drop across the capacitor, V_o. To do this you may need to redraw the circuit in Fig. 8.12 for $t < 0$, replacing the inductor with a short circuit and the capacitor with an open circuit.

2. Find the final current through the inductor, I_f and the final voltage drop across the capacitor, V_f. To do this you may need to redraw the circuit in Fig. 8.12 for $t \geq 0$, replacing the inductor with a short circuit and the capacitor with an open circuit.

3. To find α and ω_o, draw the circuit in Fig. 8.12 for $t \geq 0$. Use the values of the resistor, inductor, and capacitor and the appropriate equations for α and ω_o. Then compare the values of α^2 and ω_o^2 to determine the form of the response. Remember that the response variable will be the current in the inductor for the natural response of the series RLC circuit and for the step response of the parallel RLC circuit; the response variable will be the voltage drop across the capacitor for the natural response of the parallel RLC circuit and for the step response of the series RLC circuit.

4. Write the two equations needed to solve for the coefficients in the response from Step 3. The first equation is constructed by equating the value of the response equation at $t = 0$ with the initial condition for the voltage or current from the circuit, in Step 1. The second equation is constructed by equation the value of the first derivative of the response equation at $t = 0$ with the initial condition for the first derivative of the voltage or current in the circuit, which will be determined using additional circuit analysis. Solve the two equations and write the final form of the response.

5. If the quantity requested for the circuit in Fig. 8.12 is the same as the one you calculated in Step 4, you are done. Otherwise, use additional circuit analysis to calculate the quantity requested in Fig. 8.12.

Practice Problem 8.2

Find $i_R(t)$ for the circuit shown in Fig. 8.13.

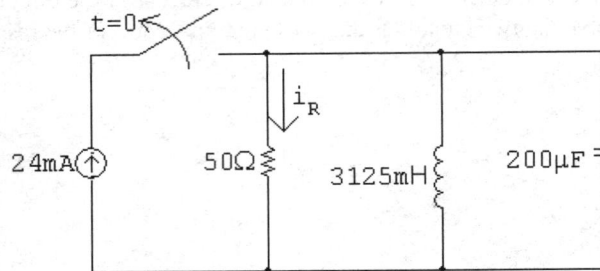

Figure 8.13: The circuit for Practice Problem 8.2.

1. Find the initial current through the inductor, I_o and the initial voltage drop across the capacitor, V_o. To do this you may need to redraw the circuit in Fig. 8.13 for $t < 0$, replacing the inductor with a short circuit and the capacitor with an open circuit.

2. Find the final current through the inductor, I_f and the final voltage drop across the capacitor, V_f. To do this you may need to redraw the circuit in Fig. 8.13 for $t \geq 0$, replacing the inductor with a short circuit and the capacitor with an open circuit.

3. To find α and ω_o, draw the circuit in Fig. 8.13 for $t \geq 0$. Use the values of the resistor, inductor, and capacitor and the appropriate equations for α and ω_o. Then compare the values of α^2 and ω_o^2 to determine the form of the response. Remember that the response variable will be the current in the inductor for the natural response of the series RLC circuit and for the step response of the parallel RLC circuit; the response variable will be the voltage drop across the capacitor for the natural response of the parallel RLC circuit and for the step response of the series RLC circuit.

4. Write the two equations needed to solve for the coefficients in the response from Step 3. The first equation is constructed by equating the value of the response equation at $t = 0$ with the initial condition for the voltage or current from the circuit, in Step 1. The second equation is constructed by equation the value of the first derivative of the response equation at $t = 0$ with the initial condition for the first derivative of the voltage or current in the circuit, which will be determined using additional circuit analysis. Solve the two equations and write the final form of the response.

5. If the quantity requested for the circuit in Fig. 8.13 is the same as the one you calculated in Step 4, you are done. Otherwise, use additional circuit analysis to calculate the quantity requested in Fig. 8.13.

Practice Problem 8.3

Find $i_L(t)$ for the circuit shown in Fig. 8.14.

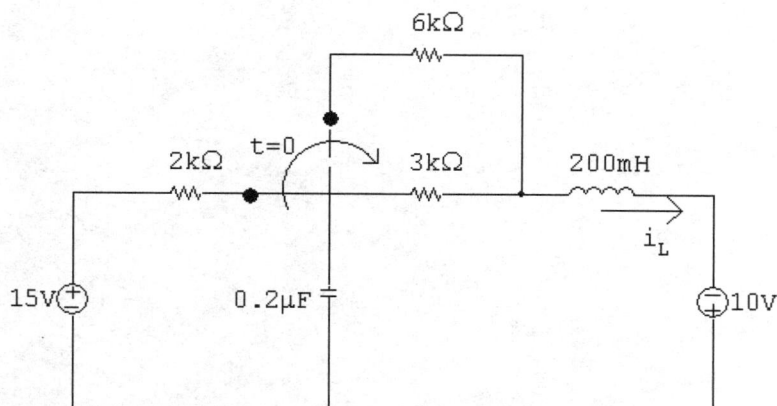

Figure 8.14: The circuit for Practice Problem 8.3.

1. Find the initial current through the inductor, I_o and the initial voltage drop across the capacitor, V_o. To do this you may need to redraw the circuit in Fig. 8.14 for $t < 0$, replacing the inductor with a short circuit and the capacitor with an open circuit.

2. Find the final current through the inductor, I_f and the final voltage drop across the capacitor, V_f. To do this you may need to redraw the circuit in Fig. 8.14 for $t \geq 0$, replacing the inductor with a short circuit and the capacitor with an open circuit.

3. To find α and ω_o, draw the circuit in Fig. 8.14 for $t \geq 0$. Use the values of the resistor, inductor, and capacitor and the appropriate equations for α and ω_o. Then compare the values of α^2 and ω_o^2 to determine the form of the response. Remember that the response variable will be the current in the inductor for the natural response of the series RLC circuit and for the step response of the parallel RLC circuit; the response variable will be the voltage drop across the capacitor for the natural response of the parallel RLC circuit and for the step response of the series RLC circuit.

4. Write the two equations needed to solve for the coefficients in the response from Step 3. The first equation is constructed by equating the value of the response equation at $t = 0$ with the initial condition for the voltage or current from the circuit, in Step 1. The second equation is constructed by equation the value of the first derivative of the response equation at $t = 0$ with the initial condition for the first derivative of the voltage or current in the circuit, which will be determined using additional circuit analysis. Solve the two equations and write the final form of the response.

5. If the quantity requested for the circuit in Fig. 8.14 is the same as the one you calculated in Step 4, you are done. Otherwise, use additional circuit analysis to calculate the quantity requested in Fig. 8.14.

Practice Problem 8.4

There is no initial energy stored in the circuit shown in Fig. 8.15. Find $v_R(t)$ for this circuit.

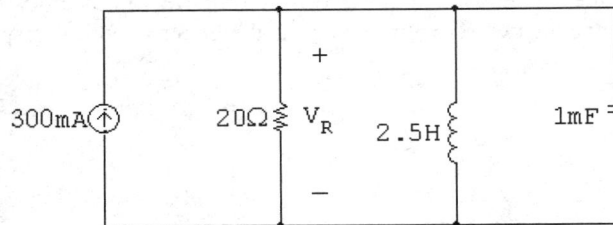

Figure 8.15: The circuit for Practice Problem 8.4.

1. Find the initial current through the inductor, I_o and the initial voltage drop across the capacitor, V_o. To do this you may need to redraw the circuit in Fig. 8.15 for $t < 0$, replacing the inductor with a short circuit and the capacitor with an open circuit.

2. Find the final current through the inductor, I_f and the final voltage drop across the capacitor, V_f. To do this you may need to redraw the circuit in Fig. 8.15 for $t \geq 0$, replacing the inductor with a short circuit and the capacitor with an open circuit.

3. To find α and ω_o, draw the circuit in Fig. 8.15 for $t \geq 0$. Use the values of the resistor, inductor, and capacitor and the appropriate equations for α and ω_o. Then compare the values of α^2 and ω_o^2 to determine the form of the response. Remember that the response variable will be the current in the inductor for the natural response of the series RLC circuit and for the step response of the parallel RLC circuit; the response variable will be the voltage drop across the capacitor for the natural response of the parallel RLC circuit and for the step response of the series RLC circuit.

4. Write the two equations needed to solve for the coefficients in the response from Step 3. The first equation is constructed by equating the value of the response equation at $t = 0$ with the initial condition for the voltage or current from the circuit, in Step 1. The second equation is constructed by equation the value of the first derivative of the response equation at $t = 0$ with the initial condition for the first derivative of the voltage or current in the circuit, which will be determined using additional circuit analysis. Solve the two equations and write the final form of the response.

5. If the quantity requested for the circuit in Fig. 8.15 is the same as the one you calculated in Step 4, you are done. Otherwise, use additional circuit analysis to calculate the quantity requested in Fig. 8.15.

Practice Problem 8.5

Find $i_L(t)$ for the circuit shown in Fig. 8.16.

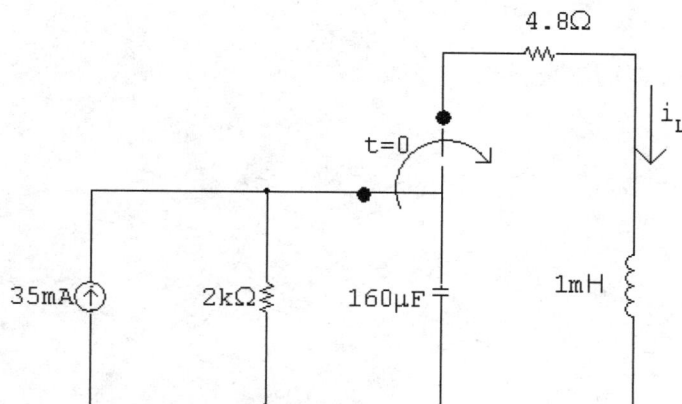

Figure 8.16: The circuit for Practice Problem 8.5.

1. Find the initial current through the inductor, I_o and the initial voltage drop across the capacitor, V_o. To do this you may need to redraw the circuit in Fig. 8.16 for $t < 0$, replacing the inductor with a short circuit and the capacitor with an open circuit.

2. Find the final current through the inductor, I_f and the final voltage drop across the capacitor, V_f. To do this you may need to redraw the circuit in Fig. 8.16 for $t \geq 0$, replacing the inductor with a short circuit and the capacitor with an open circuit.

3. To find α and ω_o, draw the circuit in Fig. 8.16 for $t \geq 0$. Use the values of the resistor, inductor, and capacitor and the appropriate equations for α and ω_o. Then compare the values of α^2 and ω_o^2 to determine the form of the response. Remember that the response variable will be the current in the inductor for the natural response of the series RLC circuit and for the step response of the parallel RLC circuit; the response variable will be the voltage drop across the capacitor for the natural response of the parallel RLC circuit and for the step response of the series RLC circuit.

4. Write the two equations needed to solve for the coefficients in the response from Step 3. The first equation is constructed by equating the value of the response equation at $t = 0$ with the initial condition for the voltage or current from the circuit, in Step 1. The second equation is constructed by equation the value of the first derivative of the response equation at $t = 0$ with the initial condition for the first derivative of the voltage or current in the circuit, which will be determined using additional circuit analysis. Solve the two equations and write the final form of the response.

5. If the quantity requested for the circuit in Fig. 8.16 is the same as the one you calculated in Step 4, you are done. Otherwise, use additional circuit analysis to calculate the quantity requested in Fig. 8.16.

Practice Problem 8.6

There is no initial energy stored in the circuit shown in Fig. 8.17. Find $i_L(t)$ for this circuit.

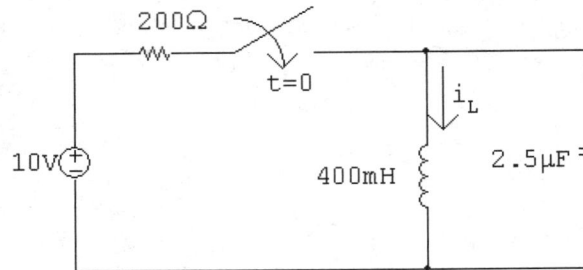

Figure 8.17: The circuit for Practice Problem 8.6.

1. Find the initial current through the inductor, I_o and the initial voltage drop across the capacitor, V_o. To do this you may need to redraw the circuit in Fig. 8.17 for $t < 0$, replacing the inductor with a short circuit and the capacitor with an open circuit.

2. Find the final current through the inductor, I_f and the final voltage drop across the capacitor, V_f. To do this you may need to redraw the circuit in Fig. 8.17 for $t \geq 0$, replacing the inductor with a short circuit and the capacitor with an open circuit.

3. To find α and ω_o, draw the circuit in Fig. 8.17 for $t \geq 0$. Use the values of the resistor, inductor, and capacitor and the appropriate equations for α and ω_o. Then compare the values of α^2 and ω_o^2 to determine the form of the response. Remember that the response variable will be the current in the inductor for the natural response of the series RLC circuit and for the step response of the parallel RLC circuit; the response variable will be the voltage drop across the capacitor for the natural response of the parallel RLC circuit and for the step response of the series RLC circuit.

4. Write the two equations needed to solve for the coefficients in the response from Step 3. The first equation is constructed by equating the value of the response equation at $t = 0$ with the initial condition for the voltage or current from the circuit, in Step 1. The second equation is constructed by equation the value of the first derivative of the response equation at $t = 0$ with the initial condition for the first derivative of the voltage or current in the circuit, which will be determined using additional circuit analysis. Solve the two equations and write the final form of the response.

5. If the quantity requested for the circuit in Fig. 8.17 is the same as the one you calculated in Step 4, you are done. Otherwise, use additional circuit analysis to calculate the quantity requested in Fig. 8.17.

Practice Problem 8.7

Find $v_R(t)$ for the circuit shown in Fig. 8.18.

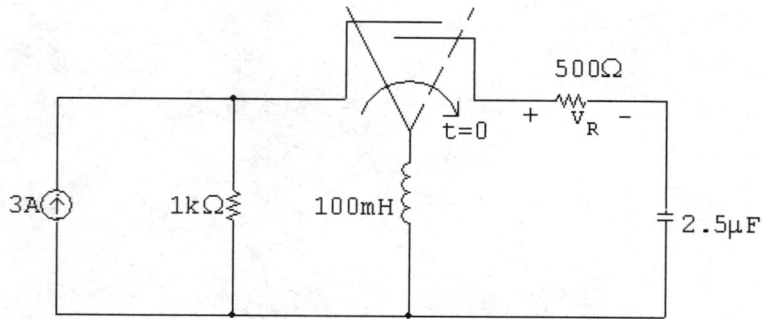

Figure 8.18: The circuit for Practice Problem 8.7.

1. Find the initial current through the inductor, I_o and the initial voltage drop across the capacitor, V_o. To do this you may need to redraw the circuit in Fig. 8.18 for $t < 0$, replacing the inductor with a short circuit and the capacitor with an open circuit.

2. Find the final current through the inductor, I_f and the final voltage drop across the capacitor, V_f. To do this you may need to redraw the circuit in Fig. 8.18 for $t \geq 0$, replacing the inductor with a short circuit and the capacitor with an open circuit.

3. To find α and ω_o, draw the circuit in Fig. 8.18 for $t \geq 0$. Use the values of the resistor, inductor, and capacitor and the appropriate equations for α and ω_o. Then compare the values of α^2 and ω_o^2 to determine the form of the response. Remember that the response variable will be the current in the inductor for the natural response of the series RLC circuit and for the step response of the parallel RLC circuit; the response variable will be the voltage drop across the capacitor for the natural response of the parallel RLC circuit and for the step response of the series RLC circuit.

4. Write the two equations needed to solve for the coefficients in the response from Step 3. The first equation is constructed by equating the value of the response equation at $t = 0$ with the initial condition for the voltage or current from the circuit, in Step 1. The second equation is constructed by equation the value of the first derivative of the response equation at $t = 0$ with the initial condition for the first derivative of the voltage or current in the circuit, which will be determined using additional circuit analysis. Solve the two equations and write the final form of the response.

5. If the quantity requested for the circuit in Fig. 8.18 is the same as the one you calculated in Step 4, you are done. Otherwise, use additional circuit analysis to calculate the quantity requested in Fig. 8.18.

Practice Problem 8.8

Find $v_L(t)$ for the circuit shown in Fig. 8.19.

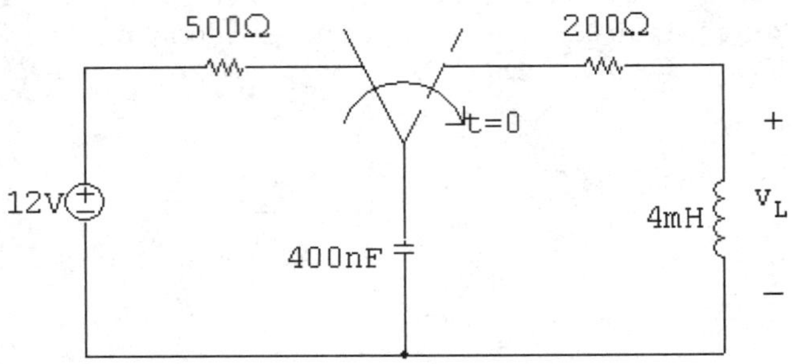

Figure 8.19: The circuit for Practice Problem 8.8.

1. Find the initial current through the inductor, I_o and the initial voltage drop across the capacitor, V_o. To do this you may need to redraw the circuit in Fig. 8.19 for $t < 0$, replacing the inductor with a short circuit and the capacitor with an open circuit.

2. Find the final current through the inductor, I_f and the final voltage drop across the capacitor, V_f. To do this you may need to redraw the circuit in Fig. 8.19 for $t \geq 0$, replacing the inductor with a short circuit and the capacitor with an open circuit.

3. To find α and ω_o, draw the circuit in Fig. 8.19 for $t \geq 0$. Use the values of the resistor, inductor, and capacitor and the appropriate equations for α and ω_o. Then compare the values of α^2 and ω_o^2 to determine the form of the response. Remember that the response variable will be the current in the inductor for the natural response of the series RLC circuit and for the step response of the parallel RLC circuit; the response variable will be the voltage drop across the capacitor for the natural response of the parallel RLC circuit and for the step response of the series RLC circuit.

4. Write the two equations needed to solve for the coefficients in the response from Step 3. The first equation is constructed by equating the value of the response equation at $t = 0$ with the initial condition for the voltage or current from the circuit, in Step 1. The second equation is constructed by equation the value of the first derivative of the response equation at $t = 0$ with the initial condition for the first derivative of the voltage or current in the circuit, which will be determined using additional circuit analysis. Solve the two equations and write the final form of the response.

5. If the quantity requested for the circuit in Fig. 8.19 is the same as the one you calculated in Step 4, you are done. Otherwise, use additional circuit analysis to calculate the quantity requested in Fig. 8.19.

Practice Problem 8.9

Find $v_R(t)$ for the circuit shown in Fig. 8.20.

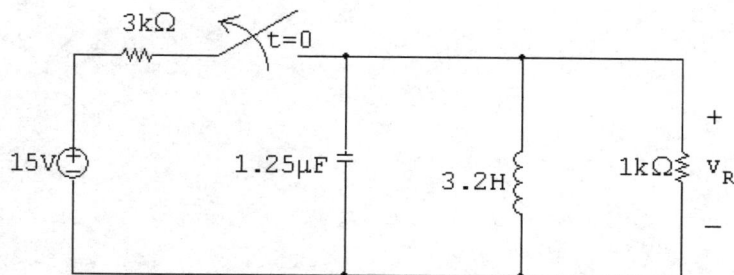

Figure 8.20: The circuit for Practice Problem 8.9.

1. Find the initial current through the inductor, I_o and the initial voltage drop across the capacitor, V_o. To do this you may need to redraw the circuit in Fig. 8.20 for $t < 0$, replacing the inductor with a short circuit and the capacitor with an open circuit.

2. Find the final current through the inductor, I_f and the final voltage drop across the capacitor, V_f. To do this you may need to redraw the circuit in Fig. 8.20 for $t \geq 0$, replacing the inductor with a short circuit and the capacitor with an open circuit.

3. To find α and ω_o, draw the circuit in Fig. 8.20 for $t \geq 0$. Use the values of the resistor, inductor, and capacitor and the appropriate equations for α and ω_o. Then compare the values of α^2 and ω_o^2 to determine the form of the response. Remember that the response variable will be the current in the inductor for the natural response of the series RLC circuit and for the step response of the parallel RLC circuit; the response variable will be the voltage drop across the capacitor for the natural response of the parallel RLC circuit and for the step response of the series RLC circuit.

4. Write the two equations needed to solve for the coefficients in the response from Step 3. The first equation is constructed by equating the value of the response equation at $t = 0$ with the initial condition for the voltage or current from the circuit, in Step 1. The second equation is constructed by equation the value of the first derivative of the response equation at $t = 0$ with the initial condition for the first derivative of the voltage or current in the circuit, which will be determined using additional circuit analysis. Solve the two equations and write the final form of the response.

5. If the quantity requested for the circuit in Fig. 8.20 is the same as the one you calculated in Step 4, you are done. Otherwise, use additional circuit analysis to calculate the quantity requested in Fig. 8.20.

Reading

- in *Electric Circuits*, ninth edition:
    - ♦ Section 8.1 — natural response of parallel RLC circuits
    - ♦ Section 8.2 — forms of the natural response
    - ♦ Section 8.3 — step response of parallel RLC circuits
    - ♦ Section 8.4 — natural and step response of series RLC circuits

Additional Problems

- 8.2 – 8.4
- 8.18 – 8.23
- 8.27
- 8.29 – 8.38
- 8.44 – 8.53
- 8.56

Solutions

- Practice Problem 8.1:

$$\alpha = 800 \text{ rad/sec}; \qquad \omega_o = 1000 \text{ rad/sec};$$
$$v_C(t) = 30 - (30 \cos 600t + 40 \sin 600t)e^{-800t} \text{ V}, \quad t \geq 0$$

- Practice Problem 8.2:

$$\alpha = 50 \text{ rad/sec}; \qquad \omega_o = 40 \text{ rad/sec};$$
$$i_R(t) = 40e^{-80t} - 40e^{-20t} \text{ mA}, \quad t \geq 0$$

- Practice Problem 8.3:

$$\alpha = 5000 \text{ rad/sec}; \qquad \omega_o = 5000 \text{ rad/sec};$$
$$i_L(t) = 5e^{-5000t} + 50{,}000te^{-5000t} \text{ mA}, \quad t \geq 0$$

- Practice Problem 8.4:

$$\alpha = 25 \text{ rad/sec}; \qquad \omega_o = 20 \text{ rad/sec};$$
$$v_R(t) = 10e^{-10t} - 10e^{-40t} \text{ V}, \quad t \geq 0$$

- Practice Problem 8.5:

$$\alpha = 2400 \text{ rad/sec}; \qquad \omega_o = 2500 \text{ rad/sec};$$
$$i_L(t) = -100 \sin 700te^{-2400t} \text{ A}, \quad t \geq 0$$

- Practice Problem 8.6:

$$\alpha = 1000 \text{ rad/sec}; \qquad \omega_o = 1000 \text{ rad/sec};$$
$$i_L(t) = 50 - 50{,}000te^{-1000t} - 50e^{-1000t} \text{ mA}, \quad t \geq 0$$

- Practice Problem 8.7:

$$\alpha = 2500 \text{ rad/sec}; \qquad \omega_o = 2000 \text{ rad/sec};$$
$$v_R(t) = 500e^{-1000t} - 2000e^{-4000t} \text{ V}, \quad t \geq 0$$

- Practice Problem 8.8:

$$\alpha = 25{,}000 \text{ rad/sec}; \qquad \omega_o = 25{,}000 \text{ rad/sec};$$

$$v_L(t) = 12e^{-25{,}000t} - 300{,}000te^{-25{,}000t} \text{ V}, \quad t \geq 0$$

- Practice Problem 8.9:

$$\alpha = 400 \text{ rad/sec}; \qquad \omega_o = 500 \text{ rad/sec};$$

$$v_R(t) = -13.33 \sin 300t\, e^{-400t} \text{ V}, \quad t \geq 0$$

Chapter 9

Sinusoidal Steady-State Circuits and AC Power

Here we analyze circuits with one or more independent sources which are **sinusoidal**. This means, for example, that if there is a voltage source in the circuit, it can be described by the equation

$$v_s(t) = V_m \cos(\omega t + \phi) \text{ V}$$

where V_m is the **amplitude** of the source in volts, ω is the **frequency** of the source in rad/sec, and ϕ is the **phase angle** of the source in degrees. When a source can be described with a sinusoid, we call it an **AC source**. The analysis technique illustrated here will produce the **steady-state** response of the circuit, not the complete response. The steady-state response is the value of the desired voltage or current that remains after the natural response has decayed to zero, as it always will. The steady-state response always has the same form as the voltage or current source. Since all of the circuits we consider have sinusoidal sources, the steady-state response will also be sinusoidal, or AC. Therefore, the circuit analysis will produce the AC steady-state response of the circuit. Remember that the AC steady-state response will be a sinusoid with the same frequency as the frequency of the voltage or current source. Therefore, in analyzing the circuits we need only calculate the magnitude and phase angle of the response, since the frequency of the response is already known.

We use the concept of a **phasor**, which is a complex number that represents the magnitude and phase angle of a sinusoid, to transform our circuit from the time domain into the frequency domain. In the frequency domain, the circuit components and their connections remain the same. Voltages and currents are replaced by phasors, and circuit component values are replaced by **impedances**. In the time domain, the equations describing the relationship between voltage and current are differential equations, but in the frequency domain the equations describing the relationship between phasor voltage and phasor current are algebraic equations. Thus, we can treat our phasor domain circuits like resistive circuits in terms of the analysis techniques. We can use the node voltage method, the mesh current method, or any other technique we choose, and the result will always be one or more algebraic equations, which we can place in standard form and solve using a calculator. The currents and voltages we solve for will be phasors.

We can check our frequency domain analysis using the concept of **complex power**, which is a complex number whose real part is **average power** and whose imaginary part is **reactive power**. We can calculate the complex power using phasor voltage and current for every component in our frequency domain circuit. If the net complex power is zero, the power in the circuit balances and the phasor voltage and current values are consistent. Then we can transform our phasor currents and voltages back into the time domain, where they describe the AC steady-state response of the circuit.

The analysis method for AC steady-state circuits can be broken into the following steps:

1. Redraw the circuit as it appears in the time domain, copying all of the components and their interconnections. This is the frequency domain circuit. For every voltage or current in the time domain that is specified as a sinusoid, label the corresponding element in the frequency domain with the phasor transform of the sinusoid. For every voltage or current in the time domain that is specified with a symbol, like v_o or i_L, label the corresponding element in the frequency domain with a phasor symbol, like $\mathbf{V}_o$ or $\mathbf{I}_L$. This includes the voltages and currents associated with dependent sources. Finally, replace the component values associated with all resistors, inductors, and capacitors in the time domain circuit with impedances in the frequency domain circuit. Remember that the impedance of a component is dependent on its time-domain value and, except for resistors, on the frequency of the source in the time domain. Use the following equations:

$$Z_{\mathrm{R}} = \mathrm{R} \qquad Z_{\mathrm{L}} = j\omega\mathrm{L} \qquad Z_{\mathrm{C}} = \frac{1}{j\omega\mathrm{C}} = \frac{-j}{\omega\mathrm{C}}$$

228

2. Choose a circuit analysis technique that suits the frequency-domain circuit. All of the analysis techniques that work for resistive circuits with DC sources can be used for the frequency domain circuit. These techniques include the node voltage method, the mesh current method, voltage and current division, source transformation, and the circuit simplification techniques like combining impedances in series and in parallel and calculating Thévenin or Norton equivalents. Using whatever circuit analysis techniques you choose, write the equations for the circuit, put them in standard form, and solve them using a calculator, a computer, or on paper. Your solution will be in the form of a voltage or current phasor.

3. Check your answer by performing an AC power balance. This means that the sum of the complex power for all components must be zero. Stated another way, this means that the sum of the average power for all components is zero and the sum of the reactive power for all components is zero. Remember that the average power for inductors and capacitors is zero, while the reactive power for resistors is zero. Independent and dependent sources can have non-zero average and reactive power. It is important to use the passive sign convention correctly when performing the power computation.

4. Inverse phasor transform your result in Step 2 to get the steady-state result in the time domain. Remember that the response will always be in the form of a sinusoid whose frequency is the same as the frequency of the voltage or current sources. The phasor result in the frequency domain specifies the amplitude and phase angle of the time domain response.

The two examples that follow illustrate the process of analyzing circuits with sinusoidal sources and calculating the steady-state response.

Example 9.1

Find the steady-state value of $v_o(t)$ for the circuit in Fig. 9.1.

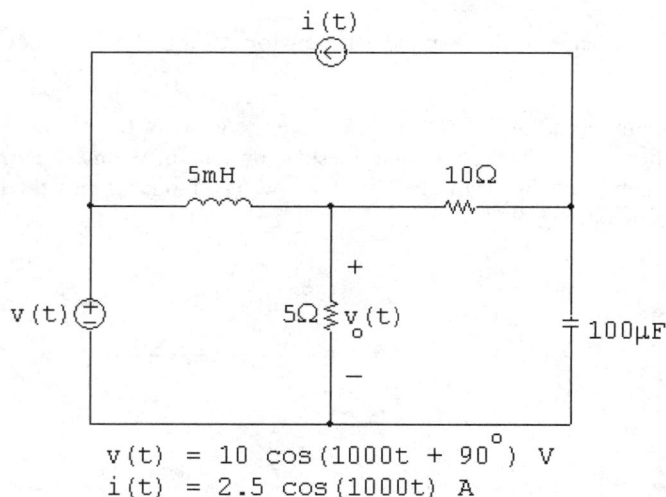

$$v(t) = 10 \cos(1000t + 90°) \text{ V}$$
$$i(t) = 2.5 \cos(1000t) \text{ A}$$

Figure 9.1: The circuit for Example 9.1

Solution

1. Redraw the circuit in Fig. 9.1. Label the voltage source in the frequency domain with the phasor transform of the sinusoid from the voltage source in the time domain:

$$\mathcal{P}\{10 \cos(1000t + 90°)\} = 10\underline{/90°} \text{ V}.$$

Then label the current source in the frequency domain with the phasor transform of the sinusoid from the current source in the time domain:

$$\mathcal{P}\{2.5 \cos 1000t\} = 2.5\underline{/0°} \text{ A}.$$

Label the output voltage v_o from the time domain circuit with a voltage phasor symbol, $\mathbf{V}_o$. Label the resistors, inductors, and capacitors with their impedances:

$$Z_{5\Omega} = 5\Omega \qquad\qquad Z_{\mathrm{L}} = j(1000)(0.005) = j5\Omega$$
$$Z_{10\Omega} = 10\Omega \qquad\qquad Z_{\mathrm{C}} = \frac{-j}{(1000)(100 \times 10^{-6})} = -j10\Omega$$

The final frequency domain circuit is shown in Fig. 9.2.

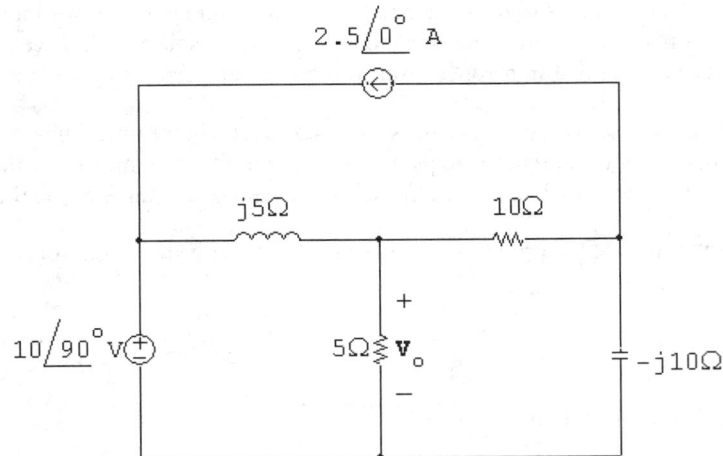

Figure 9.2: The circuit for Example 9.1, transformed into the frequency domain.

2. Analyze the frequency domain circuit in Fig. 9.2 to determine the value of the phasor $\mathbf{V}_o$. We'll illustrate by using the mesh current method for this circuit. There are three meshes, but one of the mesh currents is already specified by the current source on the perimeter of the upper mesh. Therefore, we need to label only the lower left and right meshes with phasor mesh currents, as shown in Fig. 9.3.

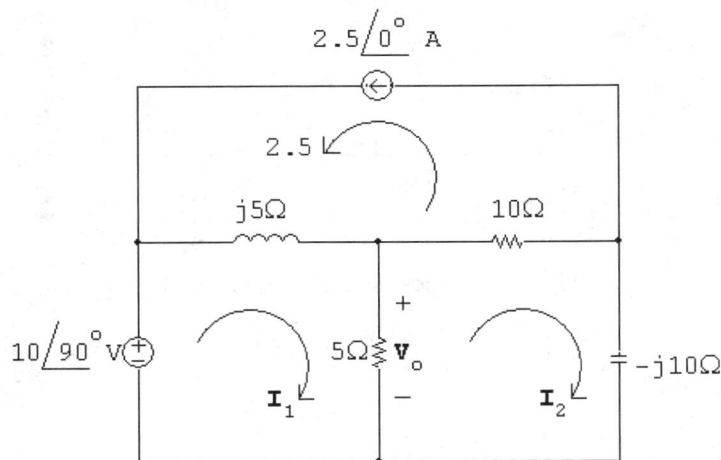

Figure 9.3: The circuit in Fig. 9.2 with mesh current phasors identified.

The mesh current equations are as follows:

Lower left mesh: $-10\underline{/90^\circ} + j5(\mathbf{I}_1 + 2.5\underline{/0^\circ}) + 5(\mathbf{I}_1 - \mathbf{I}_2) = 0$

Lower right mesh: $10(\mathbf{I}_2 + 2.5\underline{/0^\circ}) + (-j10)\mathbf{I}_2 + 5(\mathbf{I}_2 - \mathbf{I}_1) = 0$

In standard form, the equations are

$$\text{Lower left mesh:} \quad (5+j5)\mathbf{I}_1 \ + \quad (-5)\mathbf{I}_2 \quad = \quad 10\underline{/90°} - 2.5\underline{/0°}(j5)$$

$$\text{Lower right mesh:} \quad (-5)\mathbf{I}_1 \ + \ (10+5-j10)\mathbf{I}_2 \ = \quad -10(2.5\underline{/0°})$$

A calculator gives the following solutions:

$$\mathbf{I}_1 = (-1.5+j0) \text{ A}; \qquad \mathbf{I}_2 = (-1.5-j1) \text{ A}.$$

We can then use Ohm's law for impedances to calculate the desired voltage phasor:

$$\mathbf{V}_o = 5(\mathbf{I}_1 - \mathbf{I}_2) = 5(j1) = j5 = 5\underline{/90°} \text{ V}$$

3. Check the frequency domain results by performing a complex power balance. We can use the currents in the impedances, but will need both the voltage and current phasors for the sources. To get the voltage phasor for the current source, we need the voltage at the right-most node. Assuming the bottom node as the reference, this right node voltage phasor is $(-j10)\mathbf{I}_2 = (-10+j15)\text{V}$. The calculations for complex power are given below:

$$S_{\text{v.s.}} \ = \ \frac{\mathbf{VI}^*}{2} = \frac{(10\underline{/90°})(1.5+j0)}{2} = 0 + j7.5 \text{ VA}$$

$$S_{\text{c.s.}} \ = \ \frac{\mathbf{VI}^*}{2} = \frac{(-10+j15-10\underline{/90°})(2.5\underline{/0°})}{2} = -12.5 + j6.25 \text{ VA}$$

$$S_{5\Omega} \ = \ \frac{|(\mathbf{I}_1-\mathbf{I}_2)|^2(5)}{2} + j0 = \frac{|-1.5+1.5+j1|^2(5)}{2} + j0 = 2.5 + j0 \text{ VA}$$

$$S_{10\Omega} \ = \ \frac{|(\mathbf{I}_2+2.5\underline{/0°})|^2(10)}{2} + j0 = \frac{|-1.5-j1+2.5|^2(10)}{2} + j0 = 10 + j0 \text{ VA}$$

$$S_{j5} \ = \ 0+j\frac{|(\mathbf{I}_1+2.5\underline{/0°})|^2(5)}{2} = 0+j\frac{|-1.5+2.5|^2(5)}{2} = 0 + j2.5 \text{ VA}$$

$$S_{-j10} \ = \ 0+j\frac{|\mathbf{I}_2|^2(-10)}{2} = 0+j\frac{|-1.5-j1|^2(-10)}{2} = 0 - j16.25 \text{ VA}$$

We sum the complex power:

$$(0+j7.5) + (-12.5+j6.25) + (2.5+j0) + (10+j0) + (0+j2.5) + (0-j16.25) = 0 \quad \text{(checks)}$$

Thus, the phasor values we calculated are consistent with the power balance requirement.

4. Inverse phasor transform the voltage phasor $\mathbf{V}_o$ to get the time domain voltage requested originally:

$$v_o(t) = \mathcal{P}^{-1}\{5\underline{/90°}\} = 5\cos(1000t + 90°) \text{ V}.$$

Example 9.2

Find the steady-state value of $i_o(t)$ for the circuit in Fig. 9.4.

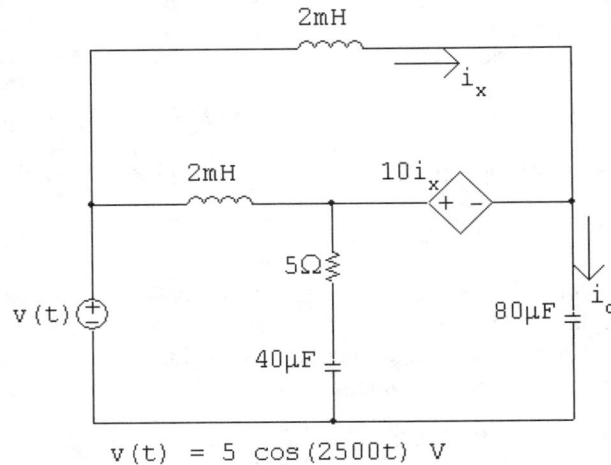

$$v(t) = 5 \cos(2500t) \text{ V}$$

Figure 9.4: The circuit for Example 9.2

Solution

1. Redraw the circuit in Fig. 9.4. Label the voltage source in the frequency domain with the phasor transform of the sinusoid from the voltage source in the time domain:

$$\mathcal{P}\{5\cos 2500t\} = 5\underline{/0°} \text{ V}.$$

Label the output current and the controlling current for the dependent source with current phasor symbols. Label the resistors, inductors, and capacitors with their impedances:

$$Z_{5\Omega} = 5\Omega$$

$$Z_\mathrm{L} = j(2500)(0.002) = j5\Omega$$

$$Z_{\mathrm{C(mid)}} = \frac{-j}{(2500)(40 \times 10^{-6})} = -j10\Omega$$

$$Z_{\mathrm{C(right)}} = \frac{-j}{(2500)(80 \times 10^{-6})} = -j5\Omega$$

The final frequency domain circuit is shown in Fig. 9.5.

2. Analyze the frequency domain circuit in Fig. 9.5 to determine the value of the phasor $\mathbf{I}_o$. We'll illustrate by using the node voltage method for this circuit. There are three non-reference essential nodes, but one of the node voltages is already specified by the voltage source on the left. Therefore, we need to label only the middle and right nodes with phasor node voltages, as shown in Fig. 9.6.

Only one node voltage equation is needed because of the supernode formed between the middle and right nodes. We need a supernode constraint equation and a dependent source constraint equation. The equations are:

$$\text{Supernode:} \quad \frac{\mathbf{V}_1 - 5}{j5} + \frac{\mathbf{V}_1}{5 - j10} + \frac{\mathbf{V}_2 - 5}{j5} + \frac{\mathbf{V}_2}{-j5} = 0$$

$$\text{Constraint:} \quad \mathbf{V}_1 - \mathbf{V}_2 = 10\,\mathbf{I}_x$$

$$\text{Constraint:} \quad \frac{5 - \mathbf{V}_2}{j5} = \mathbf{I}_x$$

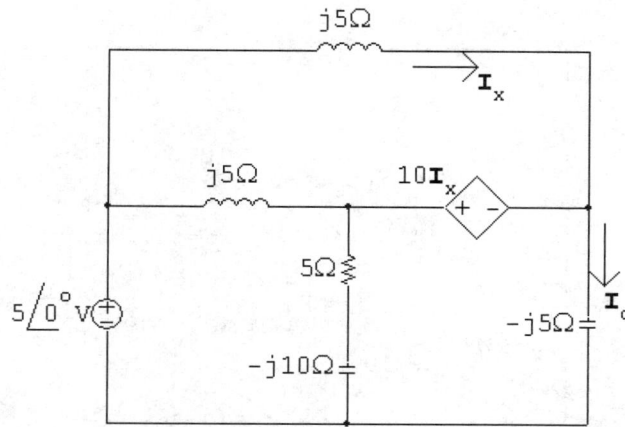

Figure 9.5: The circuit for Example 9.2, transformed into the frequency domain.

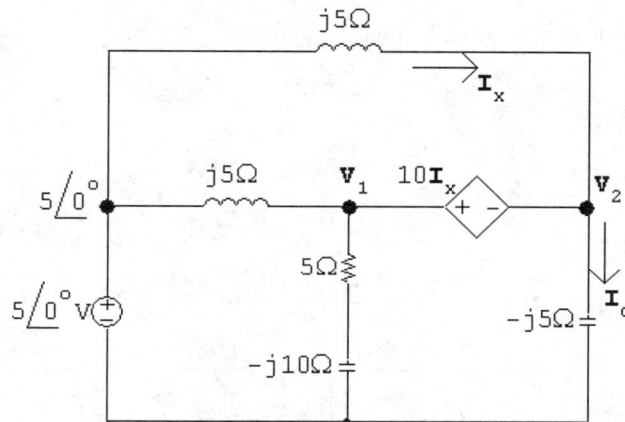

Figure 9.6: The circuit in Fig. 9.5 with node voltage phasors identified.

In standard form, the equations are

$$\text{Supernode:} \quad \left(\frac{1}{j5}+\frac{1}{5-j10}\right)\mathbf{V}_1 + \left(\frac{1}{j5}+\frac{1}{-j5}\right)\mathbf{V}_2 + (0)\mathbf{I}_x = \frac{5}{j5}+\frac{5}{j5}$$

$$\text{Constraint:} \quad (1)\mathbf{V}_1 + (-1)\mathbf{V}_2 + (-10)\mathbf{I}_x = 0$$

$$\text{Constraint:} \quad (0)\mathbf{V}_1 + \left(\frac{1}{j5}\right)\mathbf{V}_2 + (1)\mathbf{I}_x = \frac{5}{j5}$$

A calculator gives the following solutions:

$$\mathbf{V}_1 = (15 - j5)\text{ V}; \qquad \mathbf{V}_2 = (5 - j5)\text{ V}; \qquad \mathbf{I}_x = 1\text{ A}.$$

We can then use Ohm's law for impedances to calculate the desired current phasor:

$$\mathbf{I}_o = \frac{\mathbf{V}_2}{-j5} = \frac{5-j5}{-j5} = 1 + j1 = 1.414\underline{/45^\circ}\text{ A}$$

3. Check the frequency domain results by performing a complex power balance. We can use the currents in the impedances, but will need both the voltage and current phasors for the sources. We have used the node voltage values to calculate the current through every element and have labeled the circuit in Fig. 9.7 with the results.

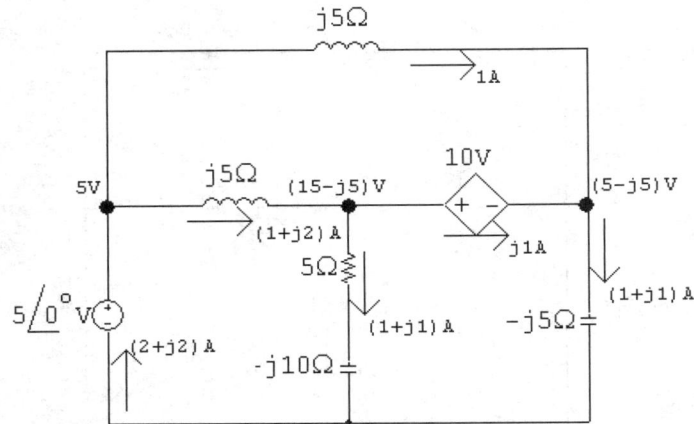

Figure 9.7: The circuit in Fig. 9.5, solved for all voltage and current phasors.

The calculations for complex power are given below:

$$S_{\text{v.s.}} = \frac{-\mathbf{V}\mathbf{I}^*}{2} = \frac{-(5)(2-j2)}{2} = -5 + j5 \text{ VA}$$

$$S_{\text{d.s.}} = \frac{\mathbf{V}\mathbf{I}^*}{2} = \frac{(10)(-j1)}{2} = 0 - j5 \text{ VA}$$

$$S_{5\Omega} = \frac{|\mathbf{I}_{5\Omega}|^2(5)}{2} + j0 = \frac{|1+j1|^2(5)}{2} + j0 = 5 + j0 \text{ VA}$$

$$S_{\text{L(top)}} = 0 + j\frac{|\mathbf{I}_{\text{L(top)}}|^2(5)}{2} = 0 + j\frac{|1|^2(5)}{2} = 0 + j2.5 \text{ VA}$$

$$S_{\text{L(mid)}} = 0 + j\frac{|\mathbf{I}_{\text{L(mid)}}|^2(5)}{2} = 0 + j\frac{|1+j2|^2(5)}{2} = 0 + j12.5 \text{ VA}$$

$$S_{\text{C(mid)}} = 0 + j\frac{|\mathbf{I}_{\text{C(mid)}}|^2(-10)}{2} = 0 + j\frac{|1+j1|^2(-10)}{2} = 0 - j10 \text{ VA}$$

$$S_{\text{C(right)}} = 0 + j\frac{|\mathbf{I}_{\text{C(right)}}|^2(-5)}{2} = 0 + j\frac{|1+j1|^2(-5)}{2} = 0 - j5 \text{ VA}$$

We sum the complex power:

$$(-5+j5) + (0-j5) + (5+j0) + (0+j2.5) + (0+j12.5) + (0-j10) + (0-j5) = 0 \quad \text{(checks)}$$

Thus, the phasor values we calculated are consistent with the power balance requirement.

4. Inverse phasor transform the voltage phasor $\mathbf{I}_o$ to get the time domain voltage requested originally:

$$i_o(t) = \mathcal{P}^{-1}\{1.414\underline{/45°}\} = 1.414\cos(2500t + 45°) \text{ A}$$

Now try using the AC steady-state circuit analysis method for each of the practice problems below.

Practice Problem 9.1

Find the steady-state value of v_o for the circuit shown in Fig. 9.8.

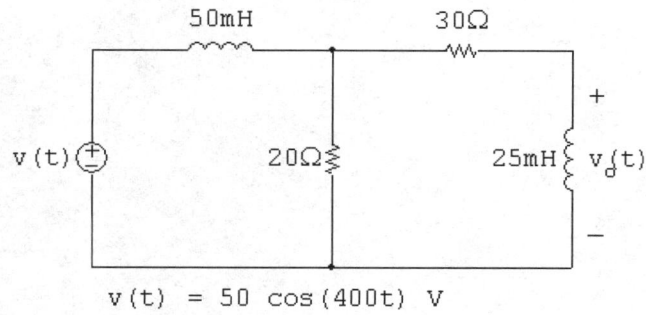

$$v(t) = 50 \cos(400t) \text{ V}$$

Figure 9.8: The circuit for Practice Problem 9.1.

1. Transform the circuit in Fig. 9.8 into the frequency domain by redrawing the circuit and replacing the voltages and currents with phasors and the resistor, inductor, and capacitor values with impedances.

2. Analyze the circuit in the frequency domain to determine the value of the phasor current $\mathbf{V}_o$.

3. Check your answer in Step 2 by performing a complex power balance.

4. Inverse phasor transform the phasor current found in Step 2 to get the steady-state value of v_o.

Practice Problem 9.2

Find the steady-state value of v_o for the circuit shown in Fig. 9.9.

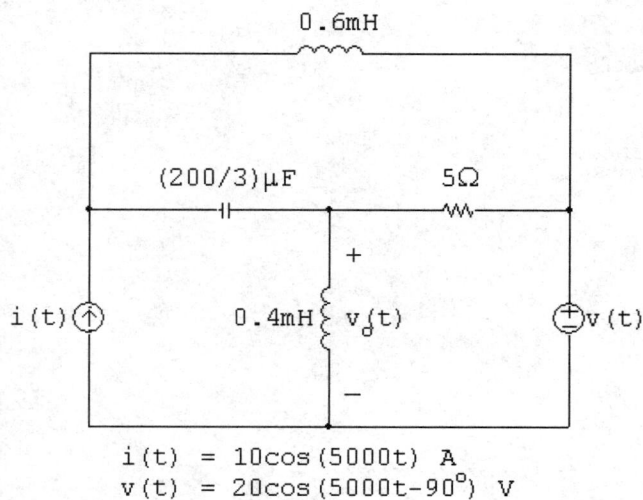

0.6mH

(200/3)μF   5Ω

i(t)   0.4mH v_o(t)   v(t)

+

−

$$i(t) = 10\cos(5000t) \text{ A}$$
$$v(t) = 20\cos(5000t-90°) \text{ V}$$

Figure 9.9: The circuit for Practice Problem 9.2.

1. Transform the circuit in Fig. 9.9 into the frequency domain by redrawing the circuit and replacing the voltages and currents with phasors and the resistor, inductor, and capacitor values with impedances.

2. Analyze the circuit in the frequency domain to determine the value of the phasor current $\mathbf{V}_o$.

3. Check your answer in Step 2 by performing a complex power balance.

4. Inverse phasor transform the phasor current found in Step 2 to get the steady-state value of v_o.

Practice Problem 9.3

Find the steady-state value of i_o for the circuit shown in Fig. 9.10.

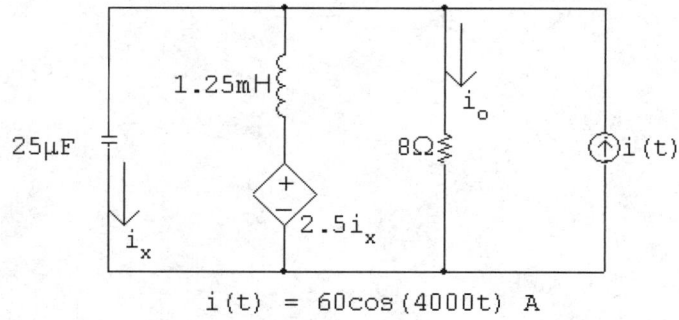

$$i(t) = 60\cos(4000t)\ \text{A}$$

Figure 9.10: The circuit for Practice Problem 9.3.

1. Transform the circuit in Fig. 9.10 into the frequency domain by redrawing the circuit and replacing the voltages and currents with phasors and the resistor, inductor, and capacitor values with impedances.

2. Analyze the circuit in the frequency domain to determine the value of the phasor current $\mathbf{I}_o$.

3. Check your answer in Step 2 by performing a complex power balance.

4. Inverse phasor transform the phasor current found in Step 2 to get the steady-state value of i_o.

Practice Problem 9.4

Find the steady-state value of i_o for the circuit shown in Fig. 9.11.

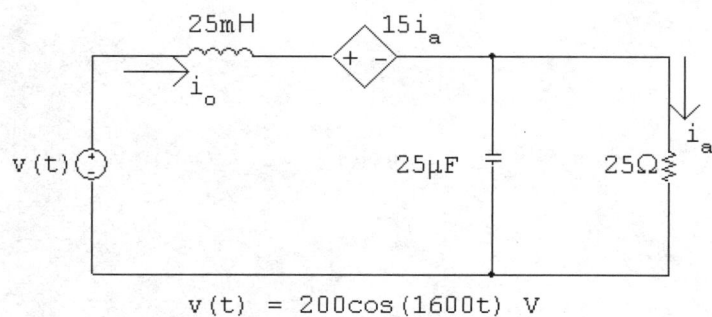

$$v(t) = 200\cos(1600t) \ V$$

Figure 9.11: The circuit for Practice Problem 9.4.

1. Transform the circuit in Fig. 9.11 into the frequency domain by redrawing the circuit and replacing the voltages and currents with phasors and the resistor, inductor, and capacitor values with impedances.

2. Analyze the circuit in the frequency domain to determine the value of the phasor current $\mathbf{I}_o$.

3. Check your answer in Step 2 by performing a complex power balance.

4. Inverse phasor transform the phasor current found in Step 2 to get the steady-state value of i_o.

Practice Problem 9.5

Find the steady-state value of v_o for the circuit shown in Fig. 9.12.

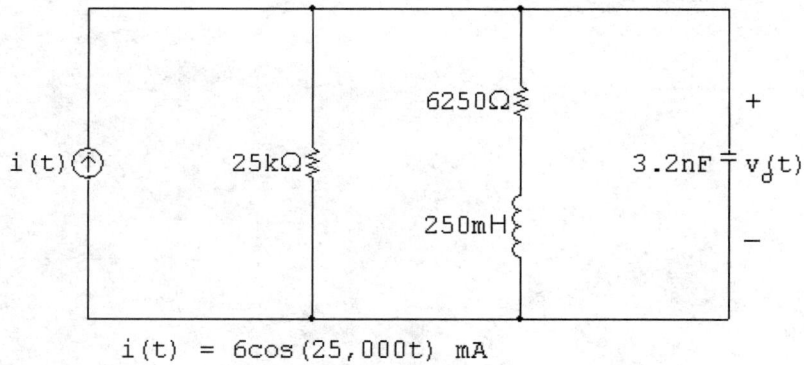

$$i(t) = 6\cos(25{,}000t) \text{ mA}$$

Figure 9.12: The circuit for Practice Problem 9.5.

1. Transform the circuit in Fig. 9.12 into the frequency domain by redrawing the circuit and replacing the voltages and currents with phasors and the resistor, inductor, and capacitor values with impedances.

2. Analyze the circuit in the frequency domain to determine the value of the phasor current $\mathbf{V}_o$.

3. Check your answer in Step 2 by performing a complex power balance.

4. Inverse phasor transform the phasor current found in Step 2 to get the steady-state value of v_o.

Practice Problem 9.6

Find the steady-state value of i_o for the circuit shown in Fig. 9.13.

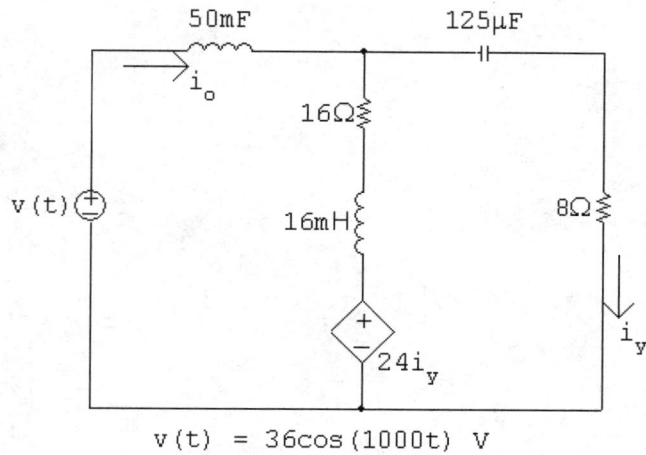

Figure 9.13: The circuit for Practice Problem 9.6.

1. Transform the circuit in Fig. 9.13 into the frequency domain by redrawing the circuit and replacing the voltages and currents with phasors and the resistor, inductor, and capacitor values with impedances.

2. Analyze the circuit in the frequency domain to determine the value of the phasor current $\mathbf{I}_o$.

3. Check your answer in Step 2 by performing a complex power balance.

4. Inverse phasor transform the phasor current found in Step 2 to get the steady-state value of i_o.

Practice Problem 9.7

Find the steady-state value of i_o for the circuit shown in Fig. 9.14.

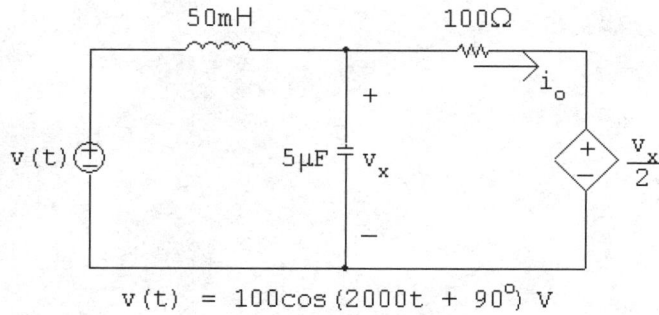

$$v(t) = 100\cos(2000t + 90°) \text{ V}$$

Figure 9.14: The circuit for Practice Problem 9.7.

1. Transform the circuit in Fig. 9.14 into the frequency domain by redrawing the circuit and replacing the voltages and currents with phasors and the resistor, inductor, and capacitor values with impedances.

2. Analyze the circuit in the frequency domain to determine the value of the phasor current $\mathbf{I}_o$.

3. Check your answer in Step 2 by performing a complex power balance.

4. Inverse phasor transform the phasor current found in Step 2 to get the steady-state value of i_o.

Practice Problem 9.8

Find the steady-state value of i_o for the circuit shown in Fig. 9.15.

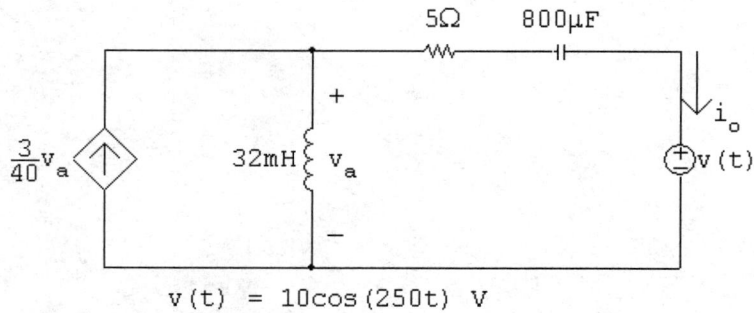

$$v(t) = 10\cos(250t) \ \text{V}$$

Figure 9.15: The circuit for Practice Problem 9.8.

1. Transform the circuit in Fig. 9.15 into the frequency domain by redrawing the circuit and replacing the voltages and currents with phasors and the resistor, inductor, and capacitor values with impedances.

2. Analyze the circuit in the frequency domain to determine the value of the phasor current $\mathbf{I}_o$.

3. Check your answer in Step 2 by performing a complex power balance.

4. Inverse phasor transform the phasor current found in Step 2 to get the steady-state value of i_o.

Reading

- in *Electric Circuits*, ninth edition:
    - ◆ Section 9.1-9.2 — sinusoids
    - ◆ Section 9.3 — phasors
    - ◆ Section 9.4 — frequency domain circuits
    - ◆ Section 9.5-9.9 — frequency domain circuit analysis
    - ◆ Section 10.1-10.6 — power in AC steady-state circuits
- Workbook section — Node Voltage Method
- Workbook section — Mesh Current Method

Additional Problems

- 9.15 – 9.16
- 9.29
- 9.31 – 9.32
- 9.44 – 9.52
- 9.54 – 9.66
- 10.5 – 10.6
- 10.16 – 10.18

Solutions

- Practice Problem 9.1:
$$v_o(t) = 7.906 \cos(400t + 18.44°) \text{ V.}$$

- Practice Problem 9.2:
$$v_o(t) = 10 \cos(5000t + 90°) \text{ V.}$$

- Practice Problem 9.3:
$$i_o(t) = 60 \cos(4000t + 53.13°) \text{ A.}$$

- Practice Problem 9.4:
$$i_o(t) = 7.071 \cos(1600t - 45°) \text{ A.}$$

- Practice Problem 9.5:
$$v_o(t) = 50 \cos(25,000t) \text{ V.}$$

- Practice Problem 9.6:
$$i_o(t) = 2 \cos(1000t - 90°) \text{ A.}$$

- Practice Problem 9.7:
$$i_o(t) = 1 \cos(2000t) \text{ A.}$$

- Practice Problem 9.8:
$$i_o(t) = 5.831 \cos(250t + 149.04°) \text{ A.}$$

Chapter 10

Laplace Transformed Circuits

In general, the describing equations for linear circuits are simultaneous differential equations. One of the most powerful mathematical techniques for solving simultaneous differential equations is the Laplace method. Here we show how to use the Laplace method to transform a circuit from the time domain to the s-domain, thereby transforming the circuit's describing differential equations into simultaneous algebraic equations which are much more manageable. The Laplace technique is also very general, in that it can be applied to circuits with any number of resistors, inductors, capacitors, and dependent sources, where the inductors and capacitors may have energy stored before the independent sources are applied. The independent sources must be able to be described by functions which have Laplace transforms. The Laplace method yields the complete response of the circuit. That is, this technique gives both the natural response of the circuit, which decays to zero in time, and the forced response, which has the same form as the independent source.

The Laplace technique begins by determining the initial conditions for the circuit. Sometimes these initial conditions are specified in the statement of the problem. Other times, the circuit contains a switch which is in one position for $t < 0$ to establish the initial conditions and in a second position for $t \geq 0$, creating the circuit meant to be analyzed. The initial conditions are represented as a voltage drop across each capacitor and the current flowing through each inductor. Then the circuit for $t \geq 0$ is Laplace transformed. This involves replacing time domain functions describing voltages and currents with their Laplace transforms and replacing each resistor, inductor, and capacitor with its Laplace transform, which in the case of inductors and capacitors involves the initial conditions. The circuit is now in the s-domain where all of the equations relating Laplace transformed voltages to Laplace transformed currents are algebraic functions of s. We can then use all of the circuit analysis techniques developed for resistive circuits with dc sources, including the node voltage method and the mesh current method. The result is an s-domain voltage or current expressed as a ratio of two polynomials in s. Once we expand this result into a sum of partial fractions, it is then easily transformed back into the time domain, where we now have a complete description of the desired voltage or current.

The Laplace transform method for circuit analysis can be broken into the following steps:

1. Determine the initial conditions for the circuit. The initial conditions represent initial energy stored in capacitors and inductors. The initial energy stored in an inductor is represented as an initial current, while the initial energy stored in a capacitor is represented as an initial voltage drop. Sometimes you will be told the values of the initial conditions in the statement of the problem, so you do not need to do any circuit analysis. In other problems you may be told that there is no initial energy stored in the circuit, so the initial value of the current in the inductors is zero as is the value of the initial voltage drop across the capacitors. Some circuits will contain a switch that is in one position for $t < 0$ while initial conditions are being established and in a second position for $t \geq 0$. When your circuit contains a switch and a dc source used to establish initial conditions you will need to do some circuit analysis to determine the value of the initial conditions. Redraw the circuit as it appears for $t < 0$. Since it is assumed that the circuit for $t < 0$ has been stable for a long time, an inductor will have been in the presence of a dc source for a long time and can be replaced with a short circuit with current I_o, while a capacitor can be replaced by an open circuit with voltage drop V_o. Use resistive circuit analysis techniques to determine the values of I_o and V_o.

2. Now consider the circuit for $t \geq 0$. We Laplace transform this circuit to get it into the s-domain. Begin by determining the Laplace transform of any voltage or current sources specified by time domain functions. To do this, use functional and operational Laplace transform tables like the ones in the text. Then Laplace transform the resistors, inductors, and capacitors in the circuit. The Laplace transform of a resistor is a resistor whose complex impedance, $Z_R(s) = R$, the resistance of the resistor. The Laplace transform of an inductor is one of two equivalent circuits — an inductor whose complex impedance $Z_L(s) = sL$ either in series with a voltage source whose value is LI_o or in parallel with a current

source whose value is I_o/s. You decide which version of the Laplace transform to use based on which version will make your s-domain circuit analysis easier. The Laplace transform of a capacitor is also one of two equivalent circuits — a capacitor whose complex impedance $Z_C(s) = 1/sC$ either in series with a voltage source whose value is CV_o or in parallel with a current source whose value is V_o/s. As with the inductor, you decide which version of the capacitor's Laplace transform to use based on which version will make your s-domain circuit analysis easier. It is important to specify the correct polarity for the sources used to represent the initial conditions for inductors and capacitors. A table is provided in the text for easy reference. Remember that if the initial conditions for the inductors and capacitors are zero, the Laplace transforms of the inductors and capacitors do not contain the voltage or current sources. Finally, replace any symbols for voltage and current in the time domain circuit, like $i(t)$ and $v(t)$ with symbols for voltages and currents in the s-domain, like $I(s)$ and $V(s)$. The resulting s-domain circuit is the Laplace transform of the original time domain circuit.

3. Since Ohm's law holds in the s-domain, that is $V(s) = Z(s)I(s)$, as do Kirchhoff's laws, all of the circuit analysis techniques developed for resistive circuits with DC sources can be used in the s-domain circuit. These techniques include the node voltage method, the mesh current method, voltage and current division, source transformation, and the circuit simplification techniques like combining complex impedances in series and in parallel and calculating Thévenin or Norton equivalents. Using whatever circuit analysis techniques you choose, write the equations for the circuit and put them in standard form. These equations will be functions of the variable s. You can use your calculator to solve these equations if it can handle the solution of equations with symbols. Otherwise you can use a computer program or Cramer's method to solve the equations. Your solution will be an s-domain voltage or current expressed as a ratio of two polynomials in s.

4. Expand the result from Step 3, the ratio of two polynomials in s, as a sum of partial fractions. Some calculators are able to do this directly. If you do the partial fraction expansion by hand, begin by factoring the denominator polynomial in s. There are two types of factors — real factors and complex conjugate pairs of factors. Either type of factor can be distinct or repeated. Thus there are four possible categories of factors — real and distinct, real and repeated, complex and distinct, and complex and repeated. Each individual factor makes up the denominator for a single partial fraction. We illustrate how to calculate the numerator constant for each partial fraction in the text and also in the examples that follow. Once the result is expressed as a sum of partial fractions, we are ready to perform the inverse Laplace transform.

5. Inverse Laplace transform each of the partial fractions by using a combination of the functional and operational Laplace transform tables in the text. The result completely describes the calculated voltage and current and will be a sum of the natural response of the circuit and its forced response. This result takes the initial conditions into account automatically.

The three examples that follow illustrate the Laplace transform method of circuit analysis. One example has distinct real roots in the Laplace transform of its result; another has repeated real roots in the Laplace transform of its result; the third has complex conjugate roots in the Laplace transform of its result.

Example 10.1

Find $v_o(t)$ when $t \geq 0$ for the circuit in Fig. 10.1.

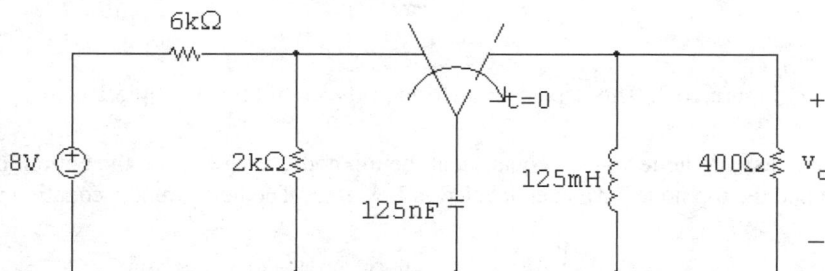

Figure 10.1: The circuit for Example 10.1

Solution

1. To determine the initial conditions, redraw the circuit in Fig. 10.1 for $t < 0$, with the switch in its left hand position. Replace the capacitor with an open circuit whose voltage drop is V_o and replace the inductor with a short circuit whose current is I_o. The resulting circuit is shown in Fig. 10.2.

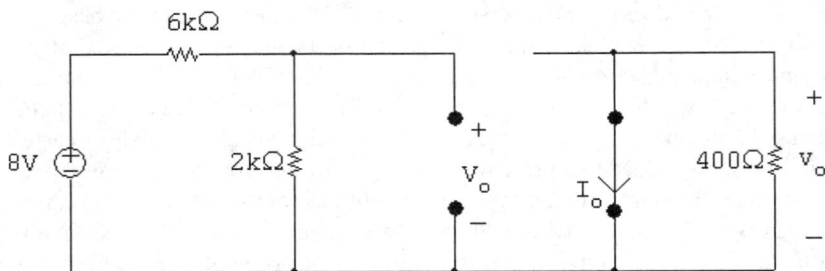

Figure 10.2: The circuit for Example 10.1, for $t < 0$.

It is easy to determine the value of I_o, since the right hand side of the circuit is open and no current can flow into it. Thus, $I_o = 0$ A. Use voltage division to determine the value of V_o, since V_o is the same as the voltage drop across the 2kΩ resistor. Thus,

$$V_o = \frac{2000}{2000 + 6000} 8 = 2 \text{ V}.$$

2. Laplace transform the circuit in Fig. 10.1 for $t \geq 0$ into the s-domain. There are no independent sources in the circuit for $t \geq 0$, so we only need to transform the resistor, inductor, and capacitor. The complex impedance of the resistor is its resistance. The complex impedance of the inductor is $sL = 0.125s$. Since the inductor's initial current is zero, there is no independent source needed to represent the initial condition. The complex impedance of the capacitor is $1/sC = 8 \times 10^6/s$. We need to incorporate the initial voltage drop across the capacitor calculated in Step 1. Here we choose the series voltage source, which has a value of $V_o/s = 2/s$ in anticipation of writing a single node voltage equation at the top node in the resulting circuit. We also replace the symbol for the output voltage in the time domain circuit, $v_o(t)$ with a symbol for the output voltage's Laplace transform, $V_o(s)$. The resulting s-domain circuit is shown in Fig. 10.3.

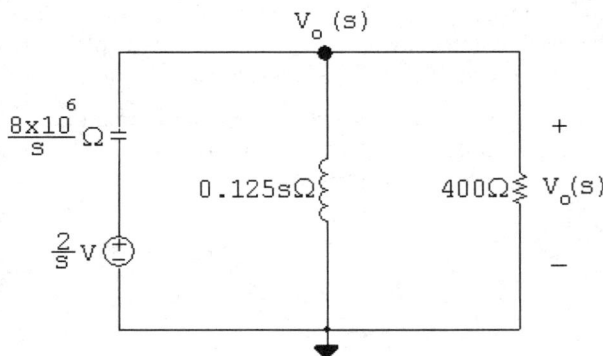

Figure 10.3: The Laplace transform of the circuit for Example 10.1.

3. Find $V_o(s)$ by writing a single node voltage equation at the top node, having made the bottom node the reference node. The reference node and the top node have been labeled in Fig. 10.3. The node voltage equation is

$$\frac{V_o(s) - \dfrac{2}{s}}{\dfrac{8 \times 10^6}{s}} + \frac{V_o(s)}{0.125s} + \frac{V_o(s)}{400} = 0$$

Solving for $V_o(s)$ we get

$$V_o(s) = \frac{2s}{s^2 + 20,000s + 64 \times 10^6}$$

This is the Laplace transform of the result we want to obtain, $v_o(t)$. Notice that we have adjusted the coefficients so that the coefficient of the highest power of s in the denominator is 1. This will allow us to factor the denominator polynomial to obtain the roots of the polynomial.

4. Factor the denominator polynomial to prepare for performing the partial fraction expansion. The result is

$$V_o(s) = \frac{2s}{(s+4000)(s+16,000)}$$

Note that in this example, the denominator factors are real and distinct. The partial fractions are in the form

$$V_o(s) = \frac{K_1}{s+4000} + \frac{K_2}{s+16,000}$$

All that remains is to calculate the values of K_1 and K_2, as shown below:

$$K_1 = (s+4000)V_o(s)|_{s=-4000} = \frac{2s}{s+16,000}\bigg|_{s=-4000} = -0.667$$

$$K_2 = (s+16,000)V_o(s)|_{s=-16,000} = \frac{2s}{s+4000}\bigg|_{s=-16,000} = 2.667$$

Therefore,

$$V_o(s) = \frac{-0.667}{s+4000} + \frac{2.667}{s+16,000}$$

5. To find $v_o(t)$, inverse Laplace transform the partial fraction expansion of $V_o(s)$:

$$v_o(t) = \mathcal{L}^{-1}\{V_o(s)\} = \mathcal{L}^{-1}\left\{\frac{-0.667}{s+4000}\right\} + \mathcal{L}^{-1}\left\{\frac{2.667}{s+16,000}\right\}$$

Using the Laplace transform tables we get

$$v_o(t) = -0.667e^{-4000t} + 2.667e^{-16,000t} \text{ V}, \quad t \geq 0.$$

Example 10.2

There is no initial energy stored in the circuit shown in Fig. 10.4. Find v_o for $t \geq 0$.

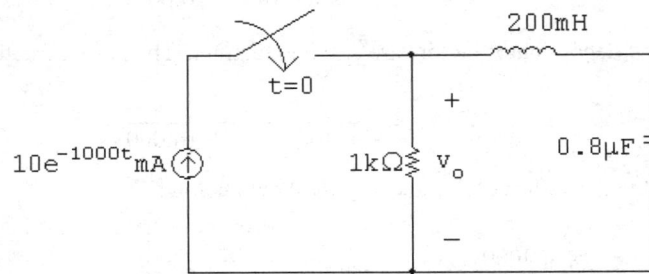

Figure 10.4: The circuit for Example 10.2

Solution

1. Since there is no initial energy stored in this circuit the initial conditions are both zero. Thus,

$$I_o = 0 \text{ A} \qquad \text{and} \qquad V_o = 0 \text{ V}.$$

2. Laplace transform the circuit in Fig. 10.4 for $t \geq 0$ into the s-domain. Replace the independent source in the circuit in Fig. 10.5 with its Laplace transform:

$$\mathcal{L}\{0.01e^{-1000t}\} = \frac{0.01}{s + 1000}$$

Transform the resistor, inductor, and capacitor. The complex impedance of the resistor is its resistance. The complex impedance of the inductor is $sL = 0.2s$. Since the inductor's initial current is zero, there is no independent source needed to represent the initial condition. The complex impedance of the capacitor is $1/sC = 125 \times 10^4/s$. Since the capacitor's initial voltage is zero, there is no independent source needed to represent the initial condition. We also replace the symbol for the output voltage in the time domain circuit, $v_o(t)$ with a symbol for the output voltage's Laplace transform, $V_o(s)$. The resulting s-domain circuit is shown in Fig. 10.5.

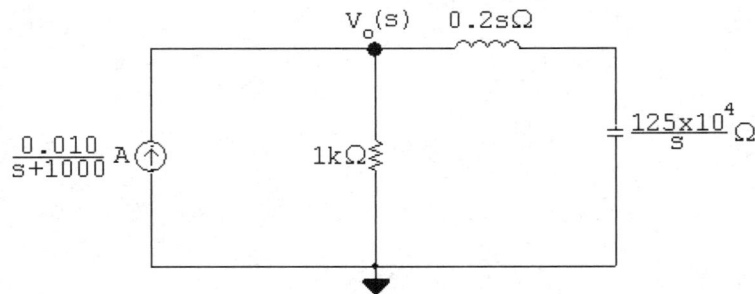

Figure 10.5: The Laplace transform of the circuit for Example 10.2.

3. Find $V_o(s)$ by writing a single node voltage equation at the top node, having made the bottom node the reference node. The reference node and the top node have been labeled in Fig. 10.5. The node voltage equation is

$$\frac{-0.01}{s + 1000} + \frac{V_o(s)}{1000} + \frac{V_o(s)}{0.2s + \dfrac{125 \times 10^4}{s}} = 0$$

Solving for $V_o(s)$ we get

$$V_o(s) = \frac{10(s^2 + 625 \times 10^4)}{(s + 1000)(s^2 + 5000s + 625 \times 10^4)}$$

This is the Laplace transform of the result we want to obtain, $v_o(t)$. Notice that we have adjusted the coefficients so that the coefficient of the highest power of s in the denominator is 1. This will allow us to factor the denominator polynomial to obtain the roots of the polynomial.

4. Factor the denominator polynomial to prepare for performing the partial fraction expansion. The result is

$$V_o(s) = \frac{10(s^2 + 625 \times 10^4)}{(s + 1000)(s + 2500)^2}$$

Note that in this example, the roots are real and repeated. The partial fractions are in the form

$$V_o(s) = \frac{K_1}{s + 1000} + \frac{K_2}{(s + 2500)^2} + \frac{K_2}{s + 2500}$$

All that remains is to calculate the values of K_1 and K_2, as shown below:

$$K_1 = (s + 1000)V_o(s)\big|_{s=-1000} = \frac{10(s^2 + 625 \times 10^4)}{s + 2500)^2}\bigg|_{s=-1000} = 32.22$$

$$K_2 = (s + 2500)^2 V_o(s)\big|_{s=-2500} = \frac{10(s^2 + 625 \times 10^4)}{(s + 1000)}\bigg|_{s=-2500} = -83,333.33$$

$$K_3 = \frac{d}{ds}\left[(s + 2500)^2 V_o(s)\right]\big|_{s=-2500} = \left[\frac{20s}{s + 1000} - \frac{10(s^2 + 625 \times 10^4)}{(s + 1000)^2}\right]\bigg|_{s=-2500}$$

$$= -22.22$$

Therefore,

$$V_o(s) = \frac{32.22}{s + 1000} + \frac{-83,333.33}{(s + 2500)^2} + \frac{-22.22}{s + 2500}$$

5. To find $v_o(t)$, inverse Laplace transform the partial fraction expansion of $V_o(s)$:

$$v_o(t) = \mathcal{L}^{-1}\{V_o(s)\} = \mathcal{L}^{-1}\left\{\frac{32.22}{s + 1000}\right\} + \mathcal{L}^{-1}\left\{\frac{-83,333.33}{(s + 2500)^2}\right\} + \mathcal{L}^{-1}\left\{\frac{-22.22}{s + 2500}\right\}$$

Thus,

$$v_o(t) = 32.22e^{-1000t} - 83,333.33te^{-2500t} - 22.22e^{-2500t} \text{ V}, \quad t \geq 0$$

Example 10.3

Find $i_o(t)$ when $t \geq 0$ for the circuit in Fig. 10.6.

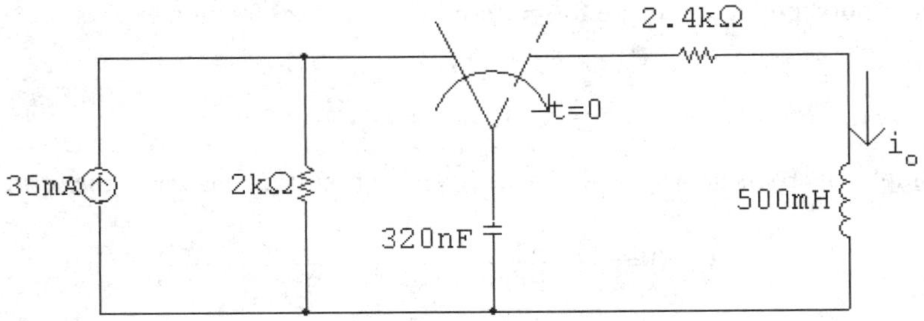

Figure 10.6: The circuit for Example 10.3

Solution

1. To determine the initial conditions, redraw the circuit in Fig. 10.6 for $t < 0$, with the switch in its left hand position. Replace the capacitor with an open circuit whose voltage drop is V_o and replace the inductor with a short circuit whose current is I_o. The resulting circuit is shown in Fig. 10.7.

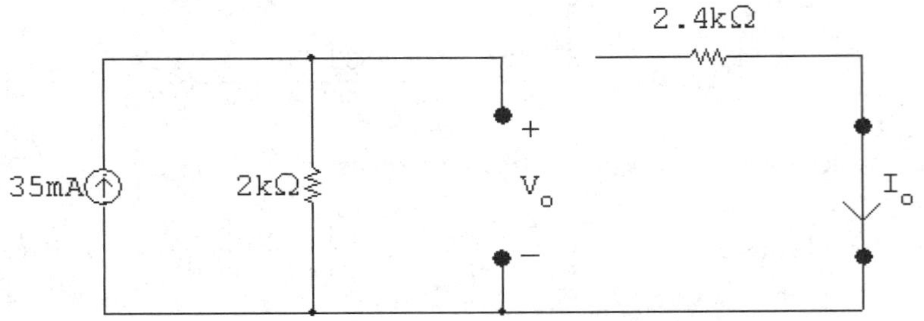

Figure 10.7: The circuit for Example 10.3, for $t < 0$.

It is easy to determine the value of I_o, since the right hand side of the circuit is open and no current can flow into it. Thus, $I_o = 0$ A. The capacitor is an open circuit so all of the current from the current source flows through the 2kΩ resistor, producing a voltage which parallel to V_o. Thus,

$$V_o = (35 \times 10^{-3})(2000) = 70 \text{ V}.$$

2. Laplace transform the circuit in Fig. 10.6 for $t \geq 0$ into the s-domain. There are no independent sources in the circuit for $t \geq 0$, so we only need to transform the resistor, inductor, and capacitor. The complex impedance of the resistor is its resistance. The complex impedance of the inductor is $sL = 0.5s$. Since the inductor's initial current is zero, there is no independent source needed to represent the initial condition. The complex impedance of the capacitor is $1/sC = 31.25 \times 10^5/s$. We need to incorporate the initial voltage drop across the capacitor calculated in Step 1. Here we choose the series voltage source, which has a value of $V_o/s = 70/s$ in anticipation of writing a single mesh current equation that incorporates all of the series connected components. We also replace the symbol for the output current in the time domain circuit, $i_o(t)$ with a symbol for the output current's Laplace transform, $I_o(s)$. The resulting s-domain circuit is shown in Fig. 10.8.

3. Find $I_o(s)$ by writing a single KVL equation around the loop of series connected components. The mesh current I_o has been labeled in Fig. 10.3. The mesh current equation is

$$\frac{-70}{s} + \frac{31.25 \times 10^5}{s} I_o + 2400 I_o + 0.5 s I_o = 0$$

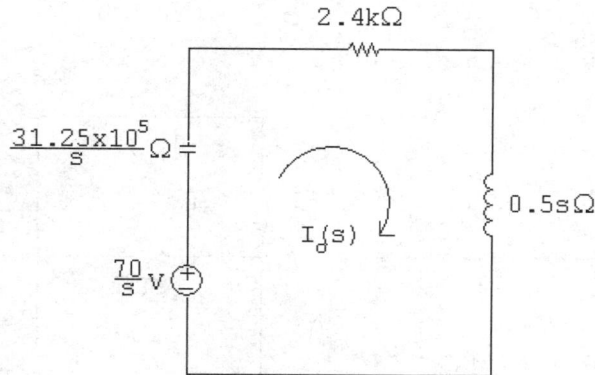

Figure 10.8: The Laplace transform of the circuit for Example 10.3.

Solving for $I_o(s)$ we get

$$I_o(s) = \frac{140}{s^2 + 4800s + 62.5 \times 10^5}$$

This is the Laplace transform of the result we want to obtain, $i_o(t)$. Notice that we have adjusted the coefficients so that the coefficient of the highest power of s in the denominator is 1. This will allow us to factor the denominator polynomial to obtain the roots of the polynomial.

4. Factor the denominator polynomial to prepare for performing the partial fraction expansion. The result is

$$I_o(s) = \frac{140}{(s + 2400 + j700)(s + 2400 - j700)}$$

Note that in this example, the denominator factors are a complex conjugate pair. The partial fractions are in the form

$$I_o(s) = \frac{K_1}{s + 2400 - j700} + \frac{K_1^*}{s + 2400 + j700}$$

All that remains is to calculate the value of K_1 as shown below:

$$
\begin{aligned}
K_1 &= (s + 2400 - j700)V_o(s)\big|_{s=-2400+j700} = \frac{140}{s + 2400 + j700}\bigg|_{s=-2400+j700} \\
&= -0.1j = 0.1\underline{/-90°}
\end{aligned}
$$

Therefore,

$$I_o(s) = \frac{0.1\underline{/-90°}}{s + 2400 - j700} + \frac{0.1\underline{/90°}}{s + 2400 + j700}$$

5. To find $i_o(t)$, inverse Laplace transform the partial fraction expansion of $I_o(s)$. When the factors of the denominator are complex conjugate pairs, the partial fractions are in the form

$$F(s) = \frac{|K|\underline{/\theta}}{s + \alpha - j\beta} + \frac{|K|\underline{/-\theta}}{s + \alpha + j\beta}$$

and the resulting inverse Laplace transform is given by

$$f(t) = 2|K|e^{-\alpha t}\cos(\beta t + \theta)$$

Therefore,

$$i_o(t) = \mathcal{L}^{-1}\{I_o(s)\} = \mathcal{L}^{-1}\left\{\frac{0.1\underline{/-90°}}{s + 2400 - j700} + \frac{0.1\underline{/90°}}{s + 2400 + j700}\right\}$$

Thus,

$$v_o(t) = 2(0.1)e^{-2400t}\cos(700t - 90°) = -0.2e^{-2400t}\sin 700t \text{ A}, \quad t \geq 0$$

Now try using the Laplace transform method for each of the practice problems below.

Practice Problem 10.1

Find the value of i_o for $t \geq 0$ in the circuit shown in Fig. 10.9.

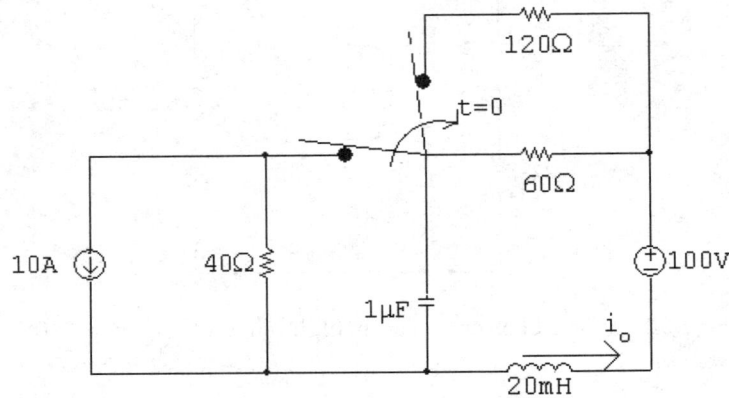

Figure 10.9: The circuit for Practice Problem 10.1.

1. Find the initial conditions for the circuit in Fig. 10.9. The value of the initial conditions may be given in the statement of the problem. If not, draw the circuit for $t < 0$, replace the capacitor with an open circuit whose voltage drop is V_o and replace the inductor with a short circuit whose current is I_o. Use resistive circuit analysis techniques to determine the values of V_o and I_o.

2. Laplace transform the circuit in Fig. 10.9 to get an s-domain circuit. Replace time domain voltages and currents with their Laplace transforms. Replace resistors, inductors, and capacitors with their Laplace transforms, which include independent sources to represent the initial conditions for inductors and capacitors.

3. Analyze the s-domain circuit from Step 2 to calculate $I_o(s)$. The result will be a ratio of two polynomials in s. Adjust the coefficients in the result so that the coefficient of the highest power of s in the denominator is 1.

4. Find the partial fraction expansion for the ratio of polynomials in s that is the value of $I_o(s)$ calculated in Step 3.

5. Inverse Laplace transform each of the terms in the partial fraction expansion from Step 4 to get the complete response $i_o(t)$ for $t > 0$.

Practice Problem 10.2

There is no initial energy stored in the circuit shown in Fig. 10.10. Find the value of v_o for $t \geq 0$ in this circuit.

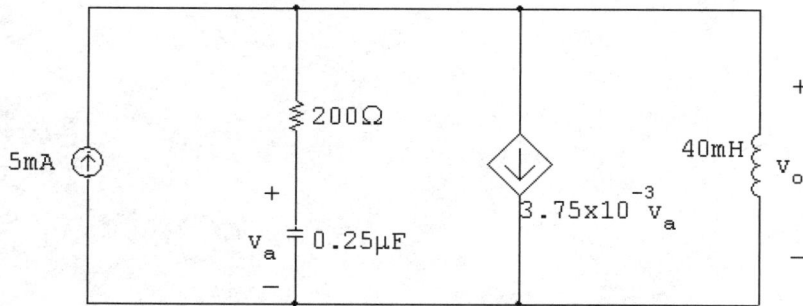

Figure 10.10: The circuit for Practice Problem 10.2.

1. Find the initial conditions for the circuit in Fig. 10.10. The value of the initial conditions may be given in the statement of the problem. If not, draw the circuit for $t < 0$, replace the capacitor with an open circuit whose voltage drop is V_o and replace the inductor with a short circuit whose current is I_o. Use resistive circuit analysis techniques to determine the values of V_o and I_o.

2. Laplace transform the circuit in Fig. 10.10 to get an s-domain circuit. Replace time domain voltages and currents with their Laplace transforms. Replace resistors, inductors, and capacitors with their Laplace transforms, which include independent sources to represent the initial conditions for inductors and capacitors.

3. Analyze the s-domain circuit from Step 2 to calculate $V_o(s)$. The result will be a ratio of two polynomials in s. Adjust the coefficients in the result so that the coefficient of the highest power of s in the denominator is 1.

4. Find the partial fraction expansion for the ratio of polynomials in s that is the value of $V_o(s)$ calculated in Step 3.

5. Inverse Laplace transform each of the terms in the partial fraction expansion from Step 4 to get the complete response $v_o(t)$ for $t > 0$.

Practice Problem 10.3

Find the value of v_o for $t \geq 0$ in the circuit shown in Fig. 10.11.

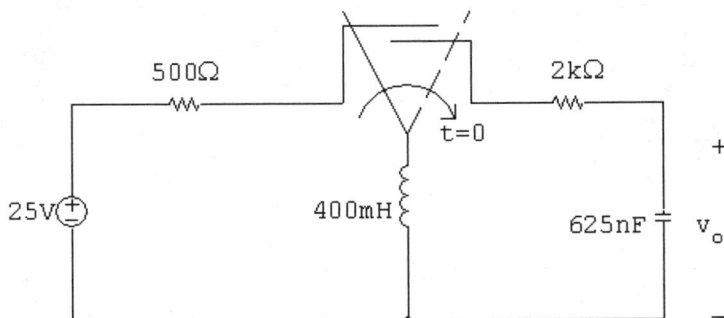

Figure 10.11: The circuit for Practice Problem 10.3.

1. Find the initial conditions for the circuit in Fig. 10.11. The value of the initial conditions may be given in the statement of the problem. If not, draw the circuit for $t < 0$, replace the capacitor with an open circuit whose voltage drop is V_o and replace the inductor with a short circuit whose current is I_o. Use resistive circuit analysis techniques to determine the values of V_o and I_o.

2. Laplace transform the circuit in Fig. 10.11 to get an s-domain circuit. Replace time domain voltages and currents with their Laplace transforms. Replace resistors, inductors, and capacitors with their Laplace transforms, which include independent sources to represent the initial conditions for inductors and capacitors.

3. Analyze the s-domain circuit from Step 2 to calculate $V_o(s)$. The result will be a ratio of two polynomials in s. Adjust the coefficients in the result so that the coefficient of the highest power of s in the denominator is 1.

4. Find the partial fraction expansion for the ratio of polynomials in s that is the value of $V_o(s)$ calculated in Step 3.

5. Inverse Laplace transform each of the terms in the partial fraction expansion from Step 4 to get the complete response $v_o(t)$ for $t > 0$.

Practice Problem 10.4

Find the value of v_o for $t \geq 0$ in the circuit shown in Fig. 10.12.

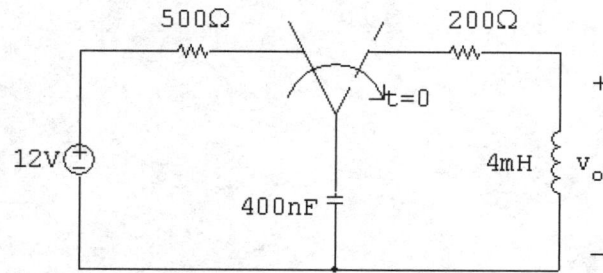

Figure 10.12: The circuit for Practice Problem 10.4.

1. Find the initial conditions for the circuit in Fig. 10.12. The value of the initial conditions may be given in the statement of the problem. If not, draw the circuit for $t < 0$, replace the capacitor with an open circuit whose voltage drop is V_o and replace the inductor with a short circuit whose current is I_o. Use resistive circuit analysis techniques to determine the values of V_o and I_o.

2. Laplace transform the circuit in Fig. 10.12 to get an s-domain circuit. Replace time domain voltages and currents with their Laplace transforms. Replace resistors, inductors, and capacitors with their Laplace transforms, which include independent sources to represent the initial conditions for inductors and capacitors.

3. Analyze the s-domain circuit from Step 2 to calculate $V_o(s)$. The result will be a ratio of two polynomials in s. Adjust the coefficients in the result so that the coefficient of the highest power of s in the denominator is 1.

4. Find the partial fraction expansion for the ratio of polynomials in s that is the value of $V_o(s)$ calculated in Step 3.

5. Inverse Laplace transform each of the terms in the partial fraction expansion from Step 4 to get the complete response $v_o(t)$ for $t > 0$.

Practice Problem 10.5

Find the value of v_o for $t \geq 0$ in the circuit shown in Fig. 10.13.

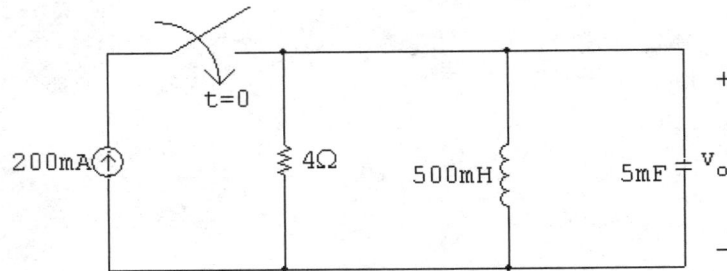

Figure 10.13: The circuit for Practice Problem 10.5.

1. Find the initial conditions for the circuit in Fig. 10.13. The value of the initial conditions may be given in the statement of the problem. If not, draw the circuit for $t < 0$, replace the capacitor with an open circuit whose voltage drop is V_o and replace the inductor with a short circuit whose current is I_o. Use resistive circuit analysis techniques to determine the values of V_o and I_o.

2. Laplace transform the circuit in Fig. 10.13 to get an s-domain circuit. Replace time domain voltages and currents with their Laplace transforms. Replace resistors, inductors, and capacitors with their Laplace transforms, which include independent sources to represent the initial conditions for inductors and capacitors.

3. Analyze the s-domain circuit from Step 2 to calculate $V_o(s)$. The result will be a ratio of two polynomials in s. Adjust the coefficients in the result so that the coefficient of the highest power of s in the denominator is 1.

4. Find the partial fraction expansion for the ratio of polynomials in s that is the value of $V_o(s)$ calculated in Step 3.

5. Inverse Laplace transform each of the terms in the partial fraction expansion from Step 4 to get the complete response $v_o(t)$ for $t > 0$.

Practice Problem 10.6

There is no initial energy stored in the circuit shown in Fig. 10.14. Find the value of v_o for $t \geq 0$ in this circuit.

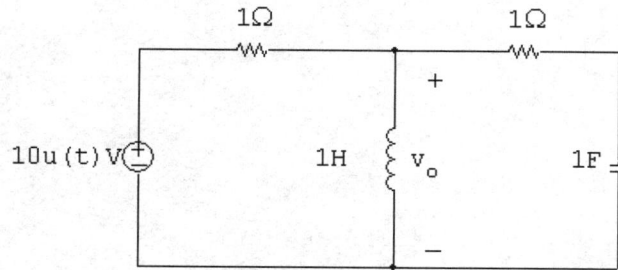

Figure 10.14: The circuit for Practice Problem 10.6.

1. Find the initial conditions for the circuit in Fig. 10.14. The value of the initial conditions may be given in the statement of the problem. If not, draw the circuit for $t < 0$, replace the capacitor with an open circuit whose voltage drop is V_o and replace the inductor with a short circuit whose current is I_o. Use resistive circuit analysis techniques to determine the values of V_o and I_o.

2. Laplace transform the circuit in Fig. 10.14 to get an s-domain circuit. Replace time domain voltages and currents with their Laplace transforms. Replace resistors, inductors, and capacitors with their Laplace transforms, which include independent sources to represent the initial conditions for inductors and capacitors.

3. Analyze the s-domain circuit from Step 2 to calculate $V_o(s)$. The result will be a ratio of two polynomials in s. Adjust the coefficients in the result so that the coefficient of the highest power of s in the denominator is 1.

4. Find the partial fraction expansion for the ratio of polynomials in s that is the value of $V_o(s)$ calculated in Step 3.

5. Inverse Laplace transform each of the terms in the partial fraction expansion from Step 4 to get the complete response $v_o(t)$ for $t > 0$.

Practice Problem 10.7

Find the value of v_o for $t \geq 0$ in the circuit shown in Fig. 10.15.

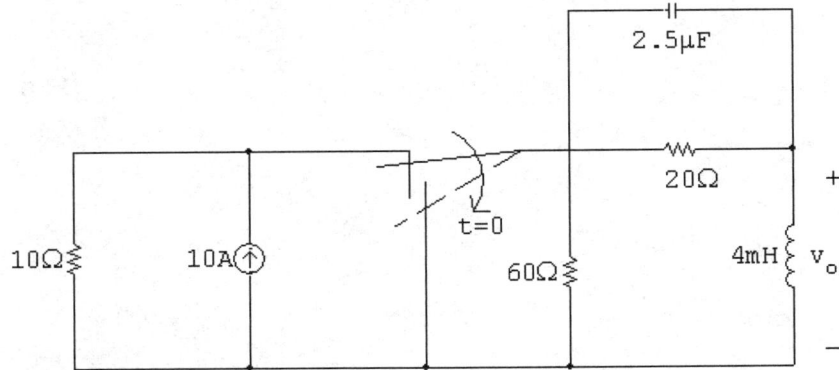

Figure 10.15: The circuit for Practice Problem 10.7.

1. Find the initial conditions for the circuit in Fig. 10.15. The value of the initial conditions may be given in the statement of the problem. If not, draw the circuit for $t < 0$, replace the capacitor with an open circuit whose voltage drop is V_o and replace the inductor with a short circuit whose current is I_o. Use resistive circuit analysis techniques to determine the values of V_o and I_o.

2. Laplace transform the circuit in Fig. 10.15 to get an s-domain circuit. Replace time domain voltages and currents with their Laplace transforms. Replace resistors, inductors, and capacitors with their Laplace transforms, which include independent sources to represent the initial conditions for inductors and capacitors.

3. Analyze the s-domain circuit from Step 2 to calculate $V_o(s)$. The result will be a ratio of two polynomials in s. Adjust the coefficients in the result so that the coefficient of the highest power of s in the denominator is 1.

4. Find the partial fraction expansion for the ratio of polynomials in s that is the value of $V_o(s)$ calculated in Step 3.

5. Inverse Laplace transform each of the terms in the partial fraction expansion from Step 4 to get the complete response $v_o(t)$ for $t > 0$.

Practice Problem 10.8

There is no initial energy stored in the circuit shown in Fig. 10.16. Find the value of v_o for $t \geq 0$ in this circuit.

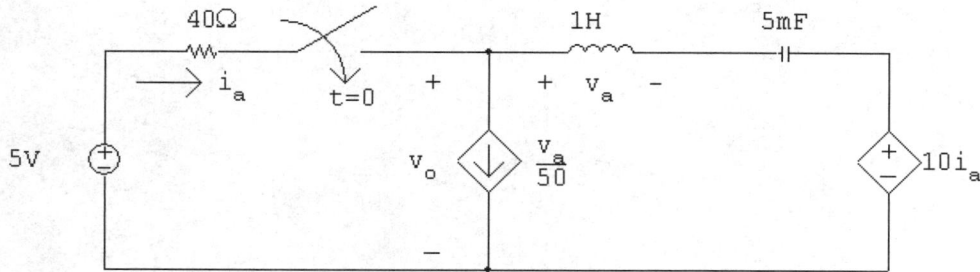

Figure 10.16: The circuit for Practice Problem 10.8.

1. Find the initial conditions for the circuit in Fig. 10.16. The value of the initial conditions may be given in the statement of the problem. If not, draw the circuit for $t < 0$, replace the capacitor with an open circuit whose voltage drop is V_o and replace the inductor with a short circuit whose current is I_o. Use resistive circuit analysis techniques to determine the values of V_o and I_o.

2. Laplace transform the circuit in Fig. 10.16 to get an s-domain circuit. Replace time domain voltages and currents with their Laplace transforms. Replace resistors, inductors, and capacitors with their Laplace transforms, which include independent sources to represent the initial conditions for inductors and capacitors.

3. Analyze the s-domain circuit from Step 2 to calculate $V_o(s)$. The result will be a ratio of two polynomials in s. Adjust the coefficients in the result so that the coefficient of the highest power of s in the denominator is 1.

4. Find the partial fraction expansion for the ratio of polynomials in s that is the value of $V_o(s)$ calculated in Step 3.

5. Inverse Laplace transform each of the terms in the partial fraction expansion from Step 4 to get the complete response $v_o(t)$ for $t > 0$.

Practice Problem 10.9

There is no initial energy stored in the inductor shown in Fig. 10.17. The initial voltage drop across the capacitor is 60V, positive at the top. Find the value of v_o for $t \geq 0$ in this circuit.

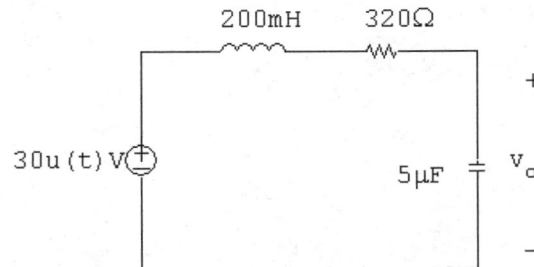

Figure 10.17: The circuit for Practice Problem 10.9.

1. Find the initial conditions for the circuit in Fig. 10.17. The value of the initial conditions may be given in the statement of the problem. If not, draw the circuit for $t < 0$, replace the capacitor with an open circuit whose voltage drop is V_o and replace the inductor with a short circuit whose current is I_o. Use resistive circuit analysis techniques to determine the values of V_o and I_o.

2. Laplace transform the circuit in Fig. 10.17 to get an s-domain circuit. Replace time domain voltages and currents with their Laplace transforms. Replace resistors, inductors, and capacitors with their Laplace transforms, which include independent sources to represent the initial conditions for inductors and capacitors.

3. Analyze the s-domain circuit from Step 2 to calculate $V_o(s)$. The result will be a ratio of two polynomials in s. Adjust the coefficients in the result so that the coefficient of the highest power of s in the denominator is 1.

4. Find the partial fraction expansion for the ratio of polynomials in s that is the value of $V_o(s)$ calculated in Step 3.

5. Inverse Laplace transform each of the terms in the partial fraction expansion from Step 4 to get the complete response $v_o(t)$ for $t > 0$.

Reading

- in *Electric Circuits*, ninth edition:

    - ◆ Section 12.1 — definition of Laplace transform
    - ◆ Section 12.4-12.5 — functional and operational Laplace transforms
    - ◆ Section 12.6 — applying Laplace transforms to circuit equations
    - ◆ Section 12.7 — inverse Laplace transforms
    - ◆ Section 13.1-13.2 — Laplace transforming circuits
    - ◆ Section 13.3 — examples of Laplace transform method

- Workbook section — Node Voltage Method

- Workbook section — Mesh Current Method

Additional Problems

- 13.9 – 13.22

- 13.25 – 13.26

- 13.28 – 13.29

- 13.31 – 13.32

Solutions

- Practice Problem 10.1:
$$i_o(t) = 5.2e^{-1000t}\cos(7000t - 15.95°) \text{ A}.$$

- Practice Problem 10.2:
$$v_o(t) = 10^4 t e^{-10,000t} + e^{-10,000t} \text{ V}.$$

- Practice Problem 10.3:
$$v_o(t) = -26.667e^{-1000t} + 26.667e^{-4000t} \text{ V}.$$

- Practice Problem 10.4:
$$v_o(t) = -300,000t e^{-25,000t} + 12e^{-25,000t} \text{ V}.$$

- Practice Problem 10.5:
$$v_o(t) = 1.333e^{-10t} - 1.333e^{-40t} \text{ V}.$$

- Practice Problem 10.6:
$$v_o(t) = 5e^{-t/2}(\cos 0.5t + \sin 0.5t) \text{ V}.$$

- Practice Problem 10.7:
$$v_o(t) = -6 \times 10^5 t e^{-10,000t} - 60e^{-10,000t} \text{ V}.$$

- Practice Problem 10.8:
$$v_o(t) = 5 - 6e^{-5t} + 4e^{-20t} \text{ V}.$$

- Practice Problem 10.9:
$$v_o(t) = 30 + 50e^{-800t}\cos(600t - 53.13°) \text{ V}.$$

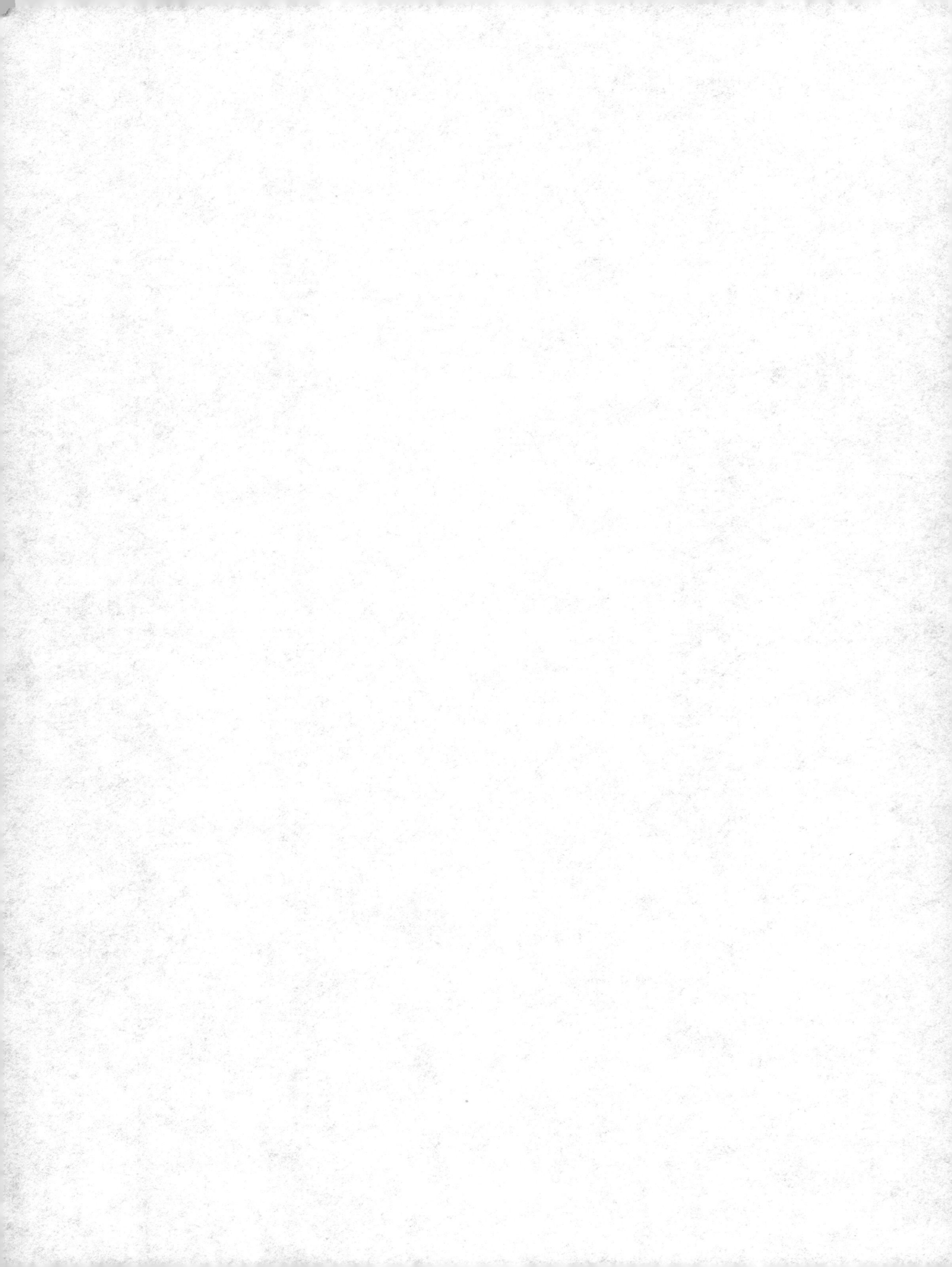